Protein Dimerization and Oligomerization in Biology

ADVANCES IN EXPERIMENTAL MEDICINE AND BIOLOGY

Recent Volumes in this Series

Volume 739
SENSING IN NATURE
Carlos López-Larrea

Volume 740
CALCIUM SIGNALING
Md. Shahidul Islam

Volume 741
STEM CELL TRANSPLANTATION
Carlos López-Larrea, Antonio López Vázquez and Beatriz Suárez Álvarez

Volume 742
ADVANCES IN MITROCHONDRIAL MEDICINE
Roberto Scatena

Volume 743
HUMAN IMMUNODEFICIENCY VIRUS TYPE 1 (HIV-1) AND BREASTFEEDING
Athena Kourtis and Marc Bulterys

Volume 744
RAMPs
William S. Spielman and Narayanan Parameswaran

Volume 745
NEW TECHNOLOGIES FOR TOXICITY TESTING
Michael Balls, Robert D. Combes and Nirmala Bhogal

Volume 746
GLIOMA: IMMUNOTHERAPEUTIC APPROACHES
Ryuya Yamanaka

Volume 747
PROTEIN DIMERIZATION AND OLIGOMERIZATION IN BIOLOGY
Jacqueline M. Matthews

Protein Dimerization and Oligomerization in Biology

Edited by

Jacqueline M. Matthews, PhD
School of Molecular Bioscience, University of Sydney, Sydney, Australia

Springer Science+Business Media, LLC

Landes Bioscience

Springer Science+Business Media, LLC
Landes Bioscience

Printed in the USA.

Springer Science+Business Media, LLC, 233 Spring Street, New York, New York 10013, USA
http://www.springer.com

Please address all inquiries to the publishers:
Landes Bioscience, 1806 Rio Grande, Austin, Texas 78701, USA
Phone: 512/ 637 6050; FAX: 512/ 637 6079
http://www.landesbioscience.com

The chapters in this book are available in the Madame Curie Bioscience Database.
http://www.landesbioscience.com/curie

Protein Dimerization and Oligomerization in Biology, edited by Jacqueline M. Matthews. Landes Bioscience / Springer Science+Business Media, LLC dual imprint / Springer series: Advances in Experimental Medicine and Biology.

ISBN: 978-1-4614-3228-9

Library of Congress Cataloging-in-Publication Data

Protein dimerization and oligomerization in biology / edited by Jacqueline M. Matthews.
p. ; cm. -- (Advances in experimental medicine and biology ; v. 747)
Includes bibliographical references and index.
Summary: "This volume has a strong focus on homo-oligomerization, which is surprisingly common. However, protein function is so often linked to both homo- and hetero-oligomerization and many heterologous interactions likely evolved from homologous interaction, so this volume also covers many aspects of hetero-oligomerization"--Provided by publisher.
ISBN 978-1-4614-3228-9 (hardback)
I. Matthews, Jacqueline M. II. Series: Advances in experimental medicine and biology ; v. 747. 0065-2598
[DNLM: 1. Protein Multimerization. 2. Protein Folding. W1 AD559 v.747 2012 / QU 55.9]
LC classification not assigned
572'.633--dc23

2012005555

PREFACE

Proteins do not act in isolation. They interact with lipids, nucleic acids, carbohydrates, small molecules and ions. And of course they interact with proteins—either like proteins (self-association/homo-oligomerization) or different proteins (heterologous association/ hetero-oligomerization). Protein-protein interactions lie at the heart of essentially all biological processes and large-scale efforts to map and characterize protein-protein interaction networks have formed a major research focus in the post-genomic era. This volume has a strong focus on homo-oligomerization, which is surprisingly common. However, protein function is so often linked to both homo- and hetero-oligomerization and many heterologous interactions likely evolved from homologous interaction, so this volume also covers many aspects of hetero-oligomerization.

Chapter 1, by Matthews and Sunde, is a fairly general overview of protein dimerization and oligomerization, covering the prevalence of homodimers and higher-order oligomers of well characterized proteins, possible origins of self-association, and some of the many functional advantages conferred by homodimers and higher order oligomers.

Traditionally, "dimerization" refers to the coming together of two similar subunits, but is often used more loosely to refer to any type of protein association—often because the stoichiometry of association is unknown. In Chapter 2, Gell, Grant and Mackay outline many of the key experimental approaches that can be used to detect protein-protein interactions and characterise the nature of protein dimerization and oligomerization. In Chapter 3, Jones describes what is known about protein association from analysis of structures, and how this information can be harnessed to predict and further analyze protein dimers and oligomers.

Enzymes form one of the best characterised class of proteins, and one in which homo-oligomerization is particularly prevalent. In Chapter 4, Mackenzie and Clarke describe the caspase system, which provides many examples of the different ways in which enzyme activity can be regulated by protein oligomerization. In Chapter 5, Griffin and Gerrard focus on the relationships between oligomeric state and enzyme function, including engineering approaches in which manipulation of oligomeric state has been used to regulate function.

Interactions between proteins and nucleic acids are essential to many aspects of cell function. In Chapter 6, Wilce, Vivian and Wilce provide a comprehensive overview of the contributions of protein dimer and oligomer formation to nucleic acid binding, while in Chapter 7, Funnell and Crossley focus on the roles that protein homo- and hetero-oligomers play in the regulation of transcription.

Many membrane channel proteins form oligomers, and in Chapter 8, Clarke and Gulbis describe, using potassium channels as an example, the intimate relationships between oligomerization and ion channel function.

One interesting mode of protein oligomerization is domain swapping—the exchange of elements of structure between like subunits. In Chapter 9, Rousseau, Schymkowitz and Itzhaki explain the implications of domain swapping in for protein folding and function, and how the same phenomenon may be involved in misfolding events.

Finally, in Chapter 10, Itzhaki and Lowe provide an overview of repeat proteins, pseudo-multimeric proteins that keep their subunits firmly in place by effectively positioning subunits on the same polypeptide chain.

Jacqueline M. Matthews, PhD

ABOUT THE EDITOR...

JACQUELINE (JACQUI) M. MATTHEWS is currently a Senior Research Fellow of the National Health and Medical Research Council of Australia, and the Professor of Protein Chemistry at the University of Sydney. Her research focuses on regulatory proteins involved in development and disease, and in particular on protein-protein and protein-DNA interactions within transcription factor complexes. Dr. Matthews received her undergraduate training (BSc Hons) in chemistry and biochemistry at the University of New South Wales in Sydney, Australia, and a PhD in biological chemistry (for work on protein folding under the supervision of Sir Professor Alan Fersht) from the University of Cambridge, UK. She is currently President of the Sydney Protein Group, a member of the Executive Council of the Protein Society and a member of the Australian Society for Biochemistry and Molecular Biology, and Australian Biophysical Society.

PARTICIPANTS

A. Clay Clark
Department of Molecular
and Structural Biochemistry
North Carolina State University
Raleigh, North Carolina
USA

Oliver B. Clarke
The Walter and Eliza Hall Institute
of Medical Research
and
Department of Medical Biology
The University of Melbourne
Melbourne, Victoria
Australia

Merlin Crossley
School of Biotechnology
and Biomolecular Sciences
University of New South Wales
and
School of Molecular Bioscience
University of Sydney
Sydney, New South Wales
Australia

Alister P. W. Funnell
School of Biotechnology
and Biomolecular Sciences
University of New South Wales
and
School of Molecular Bioscience
University of Sydney
Sydney, New South Wales
Australia

David A. Gell
School of Molecular Bioscience
University of Sydney
Sydney, New South Wales
and
Menzies Research Institute
University of Tasmania
Hobart, Tasmania
Australia

Juliet A. Gerrard
Biomolecular Interaction Centre (BIC)
School of Biological Sciences
University of Canterbury
Christchurch
New Zealand

Richard P. Grant
School of Molecular Bioscience
University of Sydney
Sydney, New South Wales
Australia

Michael D.W. Griffin
Bio21 Institute of Molecular Science
and Biotechnology
University of Melbourne
Melbourne, Victoria
Australia

Jacqueline M. Gulbis
The Walter and Eliza Hall Institute
of Medical Research
and
Department of Medical Biology
The University of Melbourne
Melbourne, Victoria
Australia

Laura S. Itzhaki
University of Cambridge
Department of Chemistry
Cambridge
UK

Susan Jones
Department of Biochemistry
School of Life Sciences
University of Sussex
Brighton
UK

Alan R. Lowe
Department of Physics
University of California-Berkeley
Berkeley, California
USA

Joel P. Mackay
School of Molecular Bioscience
University of Sydney
Sydney, New South Wales
Australia

Sarah H. MacKenzie
Department of Molecular
and Structural Biochemistry
North Carolina State University
Raleigh, North Carolina
USA

Jacqueline M. Matthews
School of Molecular Bioscience
University of Sydney
Sydney, New South Wales
Australia

Frederic Rousseau
VIB Switch Laboratory
Department of Molecular Cell Biology
Katholieke Universiteit Leuven
Leuven
Belgium

Joost Schymkowitz
VIB Switch Laboratory
Department of Molecular Cell Biology
Katholieke Universiteit Leuven
Leuven
Belgium

Margaret Sunde
School of Molecular Bioscience
and
Discipline of Pharmacology
University of Sydney
Sydney, New South Wales
Australia

Julian Vivian
Department of Biochemistry
and Molecular Biology
and
School of Biomedical Sciences
Monash Univerity
Melbourne, Victoria
Australia

Jackie Wilce
Department of Biochemistry
and Molecular Biology
and
School of Biomedical Sciences
Monash Univerity
Melbourne, Victoria
Australia

Matthew Wilce
Department of Biochemistry
and Molecular Biology
and
School of Biomedical Sciences
Monash Univerity
Melbourne, Victoria
Australia

CONTENTS

CHAPTER 1

DIMERS, OLIGOMERS, EVERYWHERE

Jacqueline M. Matthews*[,1] and Margaret Sunde[1,2]

[1]School of Molecular Bioscience, University of Sydney, Sydney, Australia; [2]Discipline of Pharmacology, University of Sydney, Sydney, Australia
**Corresponding Author: Jacqueline M. Matthews—Email: jacqui.matthews@sydney.edu.au*

Abstract: The specific self-association of proteins to form homodimers and higher order oligomers is an extremely common event in biological systems. In this chapter we review the prevalence of protein oligomerization and discuss the likely origins of this phenomenon. We also outline many of the functional advantages conferred by the dimerization or oligomerization of a wide range of different proteins and in a variety of biological roles, that are likely to have placed a selective pressure on biological systems to evolve and maintain homodimerization/oligomerization interfaces.

INTRODUCTION

Proteins rarely act alone. They commonly bind other biomolecules, including other proteins, to generate a biological response. A large percentage of proteins appear to self-associate to form dimers or higher-order oligomers. Dimerization and oligomerization can confer several different structural and functional advantages to proteins, including improved stability, regulation of activity and increased complexity.[1] Here, we consider how the phenomenon of specific self-association of proteins may have arisen and why homodimerization and homo-oligomerization of proteins remain a common feature of biological systems.

THE PREVALENCE OF PROTEIN HOMODIMERS AND HOMO-OLIGOMERS

Many proteins self-associate to form homodimers or higher order homo-oligomers (e.g., Fig. 1).[1] Anecdotally, self-association appears to be very common but it is often

Protein Dimerization and Oligomerization in Biology, edited by Jacqueline M. Matthews.

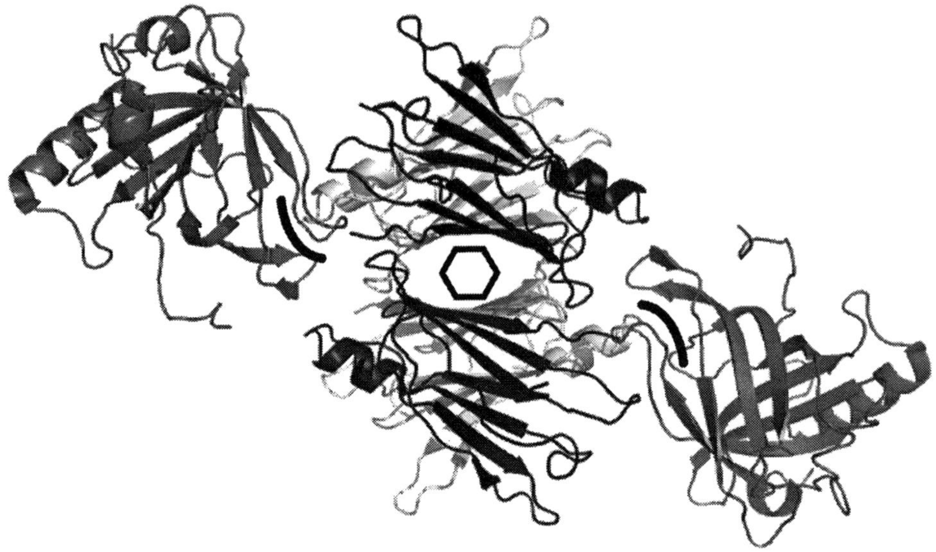

Figure 1. The Transthyretin (TTR):retinol binding protein (RBP) complex consists of a homotetramer and two heterodimers. The association of transthyretin monomers (coloured black and white) generates a homotetramer with a large central channel. The thyroid hormone thyroxine, represented as a black hexamer, binds within this channel. In addition, the tranthyretin tetramer forms a heterocomplex with two molecules of retinol binding protein (coloured in grey), which bind on either side of the tetramer. The binding site for retinol in each RBP molecule is indicated with a black line. (PDB 2WQA).

hard to quantify, partly because the stoichiometry of self-association for many proteins has not been characterized by robust biophysical methods, such as those outlined in the accompanying chapter by Gell, Grant and Mackay. Fortunately, however, several databases exist that are highly populated by very well characterized proteins, allowing us to gauge the prevalence of homodimers and higher order homo-oligomers.

The Brenda enzyme database (http://www.brenda.uni-koeln.de/) contains entries for tens of thousands of enzymes originating from all domains of life, about a third of which (~11,000 entries in July 2011) report a defined subunit composition. For this subset of enzymes, homodimers and higher order homo-oligomers by far outnumber monomers (Table 1). Not all entries have the same type of annotation so the absolute numbers vary according to how the search is carried out, but the proportions of monomer to dimer (and higher order oligomers) are all similar. Overall, monomers comprise about a quarter to a third of enzymes with a defined subunit composition and dimers plus higher order oligomers are at least twice as prevalent as monomers. Of these oligomers, dimers are most prevalent (36–38%) followed by tetramers (19%). Numbers of entries decrease rapidly as the oligomeric subunit number increases and enzymes with an odd number of subunits are less prevalent than those with an even number. Note that hetero-oligomers are a relatively poorly represented group. The database appears to be dominated by enzymes from bacterial species, but the proportions are similar for human enzymes, with the minor exception that dimers appear to be

Table 1. Subunit composition of enzymes

	Enzymes From All Species		Human Enzymes	
	Subunit Name[1]	Number of Like Subunits[2]	Subunit Name[1]	Number of Like Subunits[2]
Monomer	4847 (25%)	3641 (33%)	419 (23%)	270 (35%)
Dimer	7553 (38%)	4000 (36%)	864 (47%)	351 (45%)
Trimer	880 (4%)	382 (3%)	95 (5%)	24 (3%)
Tetramer	3719 (19%)	2075 (19%)	290 (16%)	105 (13%)
Pentamer	88 (0.4%)	42 (0.4%)	2 (0.1%)	2 (0.3%)
Hexamer	845 (4%)	449 (4%)	50 (3%)	19 (2%)
Higher order oligomers	1162 (6%)[3]	415 (4%)[4]	55 (3%)[3]	10 (1%)[4]
Hetero	620 (3%)		67 (4%)	
Total	19714	11004	1842	781

These numbers were generated through the "Search Subunits" module of the Brenda database in July 2011.
1. Subunit names as indicated were entered as the main search function. Note that this will capture both homo and heterooligomers, but heterooligomers appear to comprise a small proportion of entries.
2. The "Number of like subunits" is the output after entering "N *" (where N = 1 for monomer, 2 for dimer etc) in the Commentary window, which for many enzymes lists the number of copies of subunits. This search should exclude most hetero-oligomers, but will also exclude entries for which the subunit composition is not specified in this format (or at all in the Commentary window).
3. Subunit name: heptamer through to tetraeicosamer and poly.
4. Commentary window "N *" where N = 7–24,30,36,48,60.

particularly highly represented, apparently at the expense of higher order oligomers from tetramers upwards.

The Protein Data Bank (PDB) provides a compilation of highly characterized proteins from a much wider variety of different classes, although we note that these data are heavily biased towards soluble proteins and homomeric samples (e.g., note the relatively low number of protein hetero-oligomers; Table 2). Deposited structures are highly represented by monomers, but as at least as many proteins form dimers or higher order oligomers. Note that assignment of the biologically relevant oligomeric state from crystal structures is not trivial. It has been estimated that for 20% of dimers in the PDB the chance of misrepresentation is as high as 50%.[2] Examination of the nature and size of interfaces in crystallized complexes will reflect only the enthalpic component of complex formation and not the entropy loss on formation of the complex. Although weak interactions may be manifest in highly concentrated crystallization conditions, they may also be displaced by crystal packing contacts that result in a more favorable global energy. Several automated analysis procedures have been developed to analyze the complexes observed in crystals but complementary noncrystallographic studies should always be used to support identification of biologically significant macromolecular complexes.

In addition to the databases that report highly characterized oligomers, high-throughput studies of protein-protein interaction networks from eukaryotic organisms indicate a statistical bias towards homo-oligomeric interactions; 25–200 times more homomeric interactions were identified than could be expected if homodimers and higher order homo-oligomers randomly appeared in the course of the evolution.[3]

Table 2. Subunit composition in protein structures. Searches specified only structures that contained proteins [Macomolecule Type: Contains Protein—Yes; other options—Ignore]; and queried the [Number of Chains (Biological Assembly)] option such that monomer refers to 1, dimer to 2 etc. A 95% sequence identity cutoff was used to reduce the numbers of mutant proteins.

	One Protein Entity Only[1]	Any Number of Protein Entities[2]
Monomer	14636	14689
Dimer or larger	16780	21098
Heterooligomers[3]	-	3569
Break down of nonmonomeric structures		
Dimer	9219	10728
Trimer	2052	2561
Tetramer	3274	4151
Pentamer	133	266
Hexamer	1014	1339
Higher order oligomers[4]	1088	2053

1. [Number of Entities: Entity type—Protein; *between* 1 *and* 1 (column 1)].
2. [Number of Entities: Entity type—Protein; *between* 1 *and* 10^6 (column 2)].
3. For heterodimers [Number of Entities: Entity type—Protein; *between* 2 *and* 10^6]/The [Number of Chains (Biological Assembly): *between* 2 *and* 10^6 chains].
4. [Number of chains (Biological Entity): *between* 7 *and* 10^6 chains].

THE EVOLUTION OF PROTEIN DIMERS

Origins of Protein Self-Association

The tendency of many proteins to self-associate is a property well known to structural biologists. Indeed, for many proteins self-association is a major problem at the concentrations required for NMR spectroscopy, X-ray crystallography and techniques such as small angle scattering methodologies (e.g., ref. 4). Modeling of protein-like surfaces show they have a statistically higher affinity for self attraction compared with the propensity for attraction between *different* proteins.[5] These statistical propensities are likely to produce self-self or similar interfaces of very low affinity, but it is reasonable to assume that any such interfaces that confer a functional advantage to an organism could evolve into higher affinity interfaces that mediate specific oligomer formation. Indeed, dimer interfaces have a high degree of conservation in evolutionarily related proteins.[6]

From Simple Homo-Oligomers to Complex Systems

In prokaryotes multi-protein complexes tend to have a simpler composition than in eukayotes. For example, the catalytic core units of proteasomes are made up of two rings of alpha and two rings of beta subunits, with each ring containing seven subunits. In bacteria and archaea there is a single type of alpha and a single type of beta subunit, but in eukaryotes there are seven different types each of alpha or beta subunits (Fig. 2).[7,8]

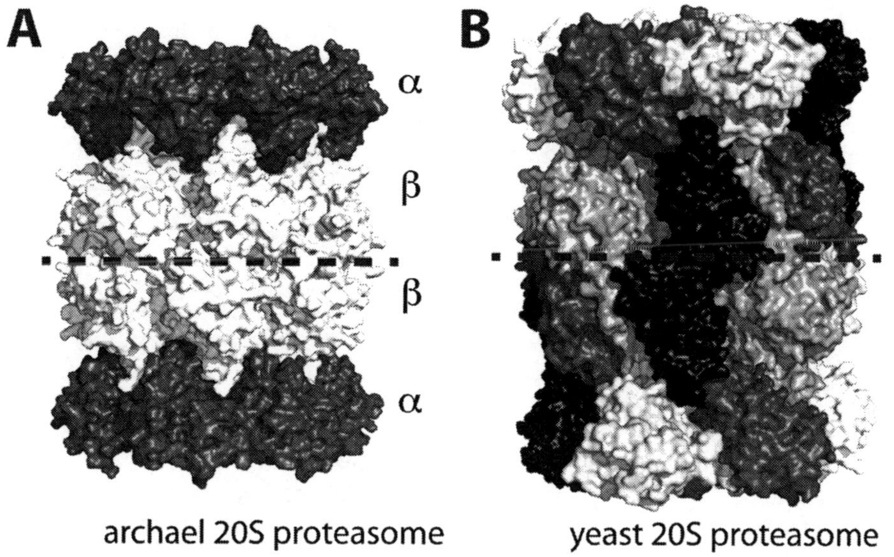

Figure 2. Comparison of the archael (*Thermoplasma acidophilum*) and yeast (*Saccharomyces cerevisiae*) 20S proteosome structures. A) The core units of the archaeal proteosome consist of two rings of alpha (α) and two rings of beta (β) subunits, with each ring containing seven subunits (alpha and beta subunits coloured white and black, respectively). B) In eukaryotes there are seven different types of alpha and seven different types of beta subunits (coloured in different shades of grey). (PDB 3IPM and 3NZJ). Black dashed lines demarcate the two β rings.

Similarly, proteasome-associated AAA ATPases tend to be homo-hexamers in bacteria and archaea and hetero-hexamers in eukaryotes. Proteins that regulate gene expression in prokaryotes are often homodimers or oligomers, but in eukaryotes, processes that regulate gene expression appear to rely heavily on the formation of multiprotein complexes.[9] The expanded sizes of eukaryotic genomes compared to prokaryotic genomes appear to have been caused in part by genome duplication events. On an evolutionary timescale identical copies of genes gradually diverge in sequence and function to form paralogs and through additional genome duplication events become families of related proteins.[10] Thus, homodimeric proteins could evolve into so called "superfamily heterodimers", families of related proteins that can form homomeric and/or heteromeric interactions with other family members. The homo- and heterodimerizing superfamilies include receptors, enzyme complexes, transcription factors and ion channels and are often functionally very important. Indeed, there is a positive correlation between the number of protein partners and importance to the viability of an organism,[11] and large scale protein-protein interaction screening studies show that proteins that can form homo-oligomers are more likely to have an increased number of binding partners.[3]

Different combinations and permutations of subunits in complexes tend to have different activities, such as transcription factor complexes targeting different DNA sequences or recruiting different cofactors (see accompanying chapters on nucleic acid binding proteins by Wilce, Vivian and Wilce and transcription factors by Funnell and Crossley). The exchange of a single component can transform a transcription complex from one that activates to one that represses transcription. This ability to use transcription factors and other regulatory proteins in a combinatorial fashion is generally thought to contribute to complexity in higher eukaryotes. In prokaryotes there appears to be some pressure to maintain a relatively small genome, which may be why these organisms depend more heavily on homomeric association. However, simpler subunit composition in multi-protein complex formation doesn't mean less regulation in prokaryotes. Rather, different mechanisms regulate the activity of oligomeric complexes, such as the regulation of protein oligomerization through posttranslational modification, small ligand- or DNA-binding.[9]

Oligomers Versus Repeat Proteins

Overall, there appears to be an evolutionary pressure placed on organisms to maintain the presence of homodimers and homo-oligomers, either directly because of the functional advantages of specific self-association, or indirectly, because proteins that can specifically self-associate are also likely to form functional hetero-oligomers. The question arises, if dimers and oligomers can confer advantages over monomeric proteins, why have they not evolved as multi-subunit single-chain proteins? This has clearly been the preferred option for a number of repeat proteins that exist as single chains, including repeats of domains that can fold independently (e.g., immunoglobulin and fibronectin domains in cytokine receptors and matrix proteins), or smaller repeated substructures (e.g., ankyrin and beta-propeller repeats, see the accompanying chapter on repeat proteins by Lowe and Itzhaki). However, in many other cases, an ability to dissociate is implicit in mechanisms that facilitate activation, deactivation or subunit exchange. In other cases the evolution of single chain multimers may be disfavoured by the need for small size and/or free termini to contribute to folding, assembly, allosteric control or interactions with cofactors.

MECHANISMS OF DIMER FORMATION

In general, the interfaces of any homomeric protein-protein interactions resemble those of heteromeric protein-protein interactions and are roughly circular and planar (see chapter by Jones on characteristics of protein interfaces). The sizes of interfaces tend to correlate with the affinity of the interaction; high affinity homo-oligomers have large interfaces and weak affinity homo-oligomers have smaller interfaces.[6,12] However, other specific modes of homodimerization exist, such as domain-swapping. Domain swapping involves the exchange of structural elements between subunits, has been involved in many structural classes and may have important implications for amyloid formation (see accompanying chapter on 3D domain-swapping by Rousseau, Schymkowitz and Itzhaki). A recently described variation on domain-swapping is the double beta-propeller structure observed in the dimer of RACK1 protein from yeast. RACK1 (the Receptor for activated C-kinase 1) is a scaffolding protein that acts as a key mediator of eukaryotic cell signaling and it is an integral part of the ribosome. As a scaffolding protein, RACK1 has many interaction partners and in some cases it appears to bind two partners simultaneously as a homodimer. A recent crystal structure suggests that a novel double beta-propeller RACK1 dimer is generated through the formation of a new, mutually shared blade 4 (Fig. 3A).[13]

An unusual metal-coordinated dimerization motif, the "zinc hook" was found in the DNA repair complex protein RAD50. A rubredoxin/zinc knuckle (CXXC, where C is Cysteine and X is any other amino acid) from each monomer coordinate a single zinc ion in a tetrahedral manner, generating a zinc finger-like coordination site (Fig. 3B).[14]

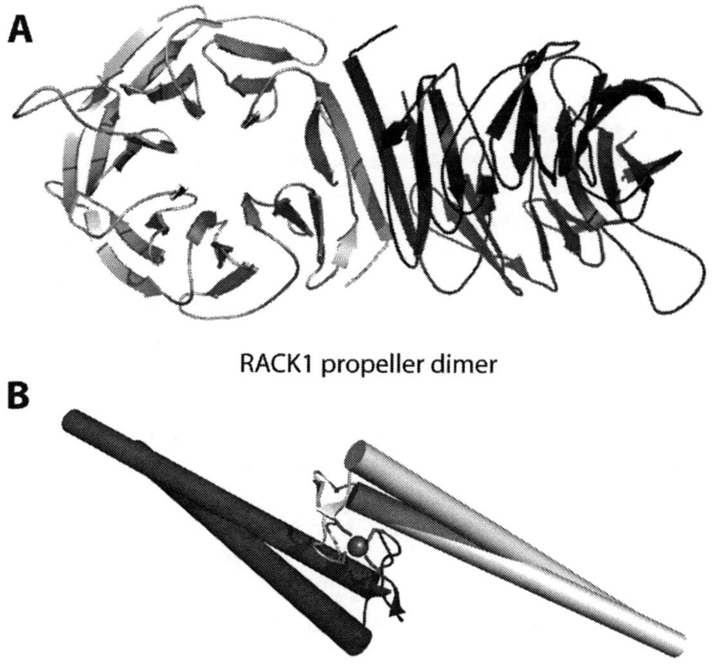

Figure 3. Novel mechanisms of dimer formation. A) The RACK1 homodimer is formed through a novel strand-swapping mechanism, where the two β-propellers interact through a shared blade. (PDB 3RFG) B) The "zinc-hook' from RAD50 comprises two rubredoxin/zinc loops that coordinate Zn(II) with tetrahedral geometry. (PDB 1L8D).

Although the zinc hook is essential for function of RAD50, a mutagenic study shows that other domains in RAD50 must also contribute to dimerization of the full length RAD50 protein.[15] Searches of the literature do not report the description of the motif in other structures, suggesting that this mode of dimerization is uncommon.

FOLDING AND ECONOMIES OF SCALE

The oligomerization of multiple, identical subunits is an economical way for organisms to form large structures. Examples of extensive networks of proteins include the long fibrous extracellular matrix proteins myosin and collagen, the assembly of hundreds or thousands of viral coat proteins into viral capsids and tubulin heterodimers (composed of α and β subunits) forming microtubules. These networks can be extremely stable (some fibrous proteins can last a human lifetime), very stable under some circumstances (viral coats are readily assembled and disassembled inside the cell but are otherwise very stable and protect the virus outside the cell), or very dynamic (tubulin subunits are added or removed from either end of microtubules in a GTP-dependent fashion, forming the ever-changing cytoskeleton that provides both a structural framework and a transport system for the cell). The assembly of multiple identical subunits into a single functional unit may help keep genome sizes small, particularly in organisms with very small or modestly sized genomes such as viruses and many prokaryotes, respectively. Small protein subunits should also fold more readily than a single large protein, reducing the load on chaperone systems within cells in order to ensure that once synthesized, proteins are correctly folded. The assembly of proteins into functional oligomers may provide an additional challenge to cells, especially for proteins that form obligate homo- or hetero-oligomers. In the cytoplasm it is thought that homo-oligomer assembly may start during translation, with protein subunits being synthesized from the same polyribosome.[16] In contrast, in the endoplasmic reticulum (where nascent proteins that are destined to reside in the plasma membrane, or be secreted from the cell, are folded and processed), oligomerization appears to occur from essentially folded monomers, but assembly can continue in the Golgi (reviewed in ref. 17). Hetero-oligomers may require more specialized assembly pathways that involve chaperones for specific subunits. For example, the hemoglobin tetramer is effectively a dimer of heterodimers and comprises two copies each of α- and β-globin. Synthesis of these two subunits is balanced such that α-globin is in slight excess. This subunit is quite prone to precipitation, but is prevented from aggregation by α-hemoglobin stabilizing protein (AHSP), which binds to and promotes folding of the α-globin monomer.[18] In contrast β-globin can form moderately stable homodimers and tetramers in the absence of its usual binding partner.

Homo-Oligomers in Protein Folding Pathways

Many proteins that regulate protein folding and unfolding form homo-oligomers.[19] For example, calreticulin and Hsp90 apparently self-associate in response to heat shock and these homo-oligomers bind misfolded or aggregating proteins with greater avidity than the monomer form.[20,21] The chaperonins, typified by the *Escherichia coli* GroEL-GroES system, are large, oligomeric ring-like assemblies. The extended interior surfaces of these rings form central cavities that protect partially unfolded proteins from aggregation and provide a favourable environment for protein folding. Similar ring-like structures are found in many of the prokaryotic chaperone-assisted proteases, such as HslU and HslUV complexes and ClpX and ClpA chaperones. However, rather than promoting folding, the central cavities of these hexameric complexes facilitate protein unfolding and proteolysis occurs.[22,23] HslU and ClpX/ClpA are AAA+ ATPases and the energy required for unfolding is thought to be provided through hydrolysis of the ATP that is bound between subunits. Almost all of the members of the AAA+ ATPase superfamily have the same ring-shaped homo-oligomeric structure and have been adapted to act as molecular machines in different cellular contexts including DNA replication, membrane fusion and peroxisome biogenesis.[24]

ACTIVATION THROUGH OLIGOMERIZATION

Interface Active Sites

The positioning of a binding site (such as an enzyme active site) at subunit interfaces provides an obvious mechanism for activation through oligomerization (e.g., Fig. 4). An increase of protein concentration above the oligomerization threshold would stimulate activation and conversely, a drop in cellular enzyme concentrations below the same

Lambda exonuclease trimer bound to DNA

Figure 4. λ exonuclease bound to DNA. The λ exonuclease binds to dsDNA ends and processively digests the 5' ended strand, generating a long 3' overhang and 5'-mononucleotides. It consists of a homotrimer with a central channel that is wide enough at one end to bind dsDNA but which narrows to allow only ssDNA through at the other end. The three subunits are illustrated in white, grey and dark grey, with the DNA oligonucleotide illustrated in black. (PDB 3SLP).

threshold would result in dissociation and inactivation. Alternatively, mechanisms such as ligand-binding or posttranslational modification (including proteolysis) could modulate oligomerization to activate or deactivate activity. In contrast, the burial of an active site at a dimer interface results in a reversal of this phenomenon, oligomerization leads to deactivation and dissociation of the oligomer leads to activation. Many examples of these mechanisms are evident in the activation and regulation of caspases, the details of which are covered in the accompanying chapter by MacKenzie and Clarke. Similar mechanisms are also used in non-enzymatic proteins, including proteins involved in subceullular localization. For example, nuclear export often involves recognition and binding of short leucine-rich nuclear export motifs (NESs)[25,26] that may mediate both nuclear export and homo-oligomerization. Structures of NESs bound to exportin proteins show that the leucine residues from the substrate NES form a hydrophobic face on an α-helix motif that binds a groove in the exportin protein CRM1.[27,28] In the case of p53, a similarly structured NES is found at tetramer interface of the tetramerization domain, suggesting that tetramer formation masks nuclear export.[29] Leucine-rich NESs have features in common with leucine zippers/coiled-coils—repeating leucine (or other hydrophobic) residues that are predicted to form the hydrophobic face of an amphipathic α-helix. Thus, it is likely that many other proteins use the same domains to mediate nuclear export and homo-oligomerization through NESs.

Allosteric Regulation

Homodimerization or homo-oligomerization can set the scene for allostery, i.e., when one ligand can affect the binding or catalysis of another in the absence of a direct interaction.

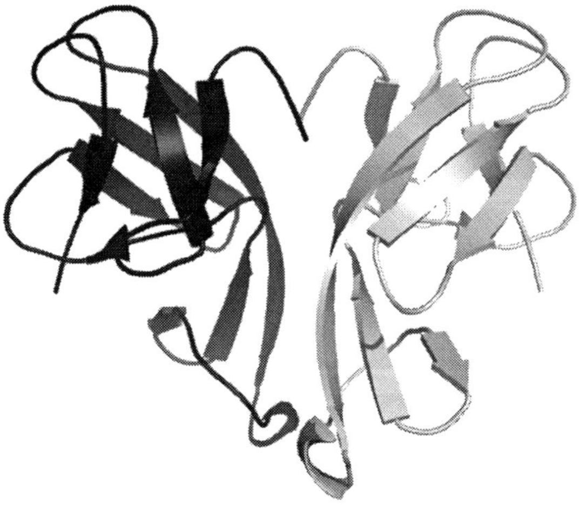

Figure 5. The methyltransferase SET domain from the *Paramecium bursaria* chlorella virus. This enzyme is a functional dimer. One site is active at a time and the dimeric nature of the enzyme gives it a higher efficiency than the monomeric form because the enzyme remains continuously associated with chromatin while methylating H3K27 and "walks" along the chromatin. (PDB 3KMA).

The concept was developed as early as 1935 to explain the positive cooperativity observed for the binding of oxygen molecules in haemoglobin,[30] although the term "allosteric" was not coined until 1961 when Monod and Jacob used the term to describe the possible regulation of the enzyme, L-threonine deaminase.[31] In recent years, advances in experimental techniques, such as increased sensitivity and resolution in nuclear magnetic resonance spectroscopy and computational power in molecular dynamics simulations, has triggered a resurgence of interest in the allostery of homomeric proteins. These new studies are being used to determine the mechanisms by which allostery is achieved in homomeric proteins.[32] The methyltransferase SET domain from the *Paramecium bursaria* chlorella virus suppresses genome-wide host transcription by methylating histone H3 at lysine 27. Unlike mammalian lysine methyltransferases, this viral SET domain functions as a dimeric enzyme (Fig. 5), remains associated with chromatin all the time while methylating H3K27 through alternate active sites and "walks" along the chromatin.[33] The dimeric nature of the enzyme allows it to have a higher efficiency than it would if it was constantly dissociating and reassociating with chromatin as a monomer. For a more comprehensive discussion on mechanisms by which protein dimerization and oligomerization is used to regulate enzyme activity see the accompanying chapter by Griffin and Gerrard.

DIMERS AND OLIGOMERS IN MEMBRANE PROTEINS

Cytokine Signaling

Cytokines and protein hormones, such as growth hormone (GH) and insulin, are extracellular proteins that signal across cell membranes by binding to the extracellular domains on membrane bound receptors. The cytokines fall into three main classes: interleukins, tumor necrosis factors (TNFs) and chemokines. Cytokine signalling provides many examples of how protein dimerization contributes to signaling across membranes.

The interleukins play an important part in the immune system and blood cell development, but have roles in many other cell types. A large subclass of interleukins, GH and similar protein hormones, share a four helical bundle core structure and a common mechanism for signalling. The receptors for these ligands contain a single transmembrane segment, the N-terminal extracellular domains contain at least one ligand binding site comprising two fibronectin Type III modules and the cytoplasmic domains either contain intrinsic tyrosine kinase activity, or recruit JAK (Janus Kinase) tyrosine kinases and trigger STAT (Signal Transducer and Activator of Transcription) signaling pathways. For many years it was thought that binding of the ligands induced dimerization of the receptors (either homodimerization or recruitment of several different receptor molecules, depending on the ligand) to trigger a cascade of intracellular signaling events, including reversible phosphorylation of receptor subunits and associated signaling molecules. This induced dimerization model of the initiation of signaling was based largely on experiments that used only the extracellular domains of the receptors to characterize ligand-receptor binding. More recently, however, it has been shown that regions from the transmembrane and intracellular domains of the receptors can be involved in dimerization and many receptors including growth hormone (GH), erythropoietin and GM-CSF receptors can exist as preformed dimers in the absence of the ligand.[34-36] New models have been proposed that invoke rotational and translational movements between receptor subunits in response to ligand binding to initiate intracellular signaling cascades (e.g., reviewed in ref. 37).

A single GH molecule binds to two identical GH receptors in an asymmetric fashion (Fig. 6A); each receptor binds a different site on GH: one with a high and one with low binding affinity. This scenario leads to a rapidly activating receptor system, but one in which the activity is attenuated as concentrations of ligand rise and form 1:1 inactive GH-receptor complexes via the high affinity binding site rather than active 1:2 complexes.[37] Many cytokines have a far more complex receptor system, including binding through two different receptor subunits and networks of interactions between the ligands and receptors (e.g., Fig. 6B and refs. 38,39). This complexity likely leads to varied and more versatile activation profiles.

Signal Amplification through Receptor Clustering

The arrangement of molecules within the recently determined crystal structure of the GM-CSF extracellular signaling complex implies that, rather than forming discrete hexameric signaling units comprising two copies each of the ligand and two different receptors, the minimal active signaling subunit is a dodecamer—two interacting hexamers that are linked through homodimeric contacts between receptor subunits.[39] Moreover, the arrangement of the asymmetric subunits in the crystal lattice and free interaction sites at the ends of the dodecamer suggests that the dodecamer is a minimal repeating unit that can form extensive clusters of signaling molecules at the cell surface. Clustering of ligand/receptor complexes generates a high density of signaling molecules that could effectively amplify a signaling event. For example, epidermal growth factor (EGF) receptor, which is usually present at very low surface concentrations, may preassemble into dimers to bring about very rapid responses to ligand binding.[40]

Death receptors (cytokine receptors that transmit apoptotic signals) are also thought to form clusters in the cell membrane in response to ligand binding. A general mechanism

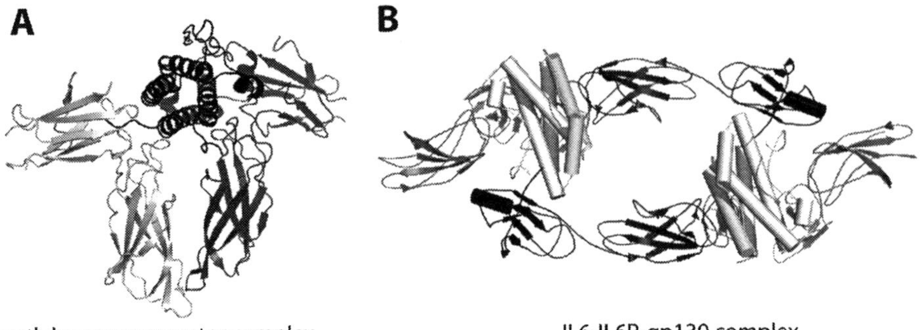

Figure 6. Cytokine and growth hormone receptor complexes. A) A single molecule of growth hormone binds asymmetrically to two identical GH receptors. There is one high affinity site on the hormone and one low affinity site, allowing for a rapidly activating system that is controlled by ligand concentration. (PDB 3HHR). B) The interleukin-6-(IL-6)interleukin-6 receptor-(IL-6R)/gp130 extracellular signaling complex is a hexameric structure. The binary IL-6-IL-6R complex forms initially. Addition of gp130 to form the ternary complex results in a cooperative and sequential transition from heterotrimer to hexamer. IL-6 is illustrated in white, domains D2 and D3 of the IL-6R in grey and domains D1, D2 and D3 of gp130 in black. (PDB 1P9M).

of clustering for this family of receptors involves the binding of specific ligands such as Fas, tumour necrosis factor (TNF) or TNF-related apoptosis-inducing ligand (TRAIL) to their cognate receptors, which triggers the generation of ceramide (a lipid found in cell membranes). Ceramide release is thought to lead to lipid raft formation which facilitates receptor clustering. TNF superfamily proteins are trimeric cytokines that each bind to three cell surface TNF receptor (TNFR) proteins (e.g., ref. 41). Some studies have shown that TNF receptor family proteins can self-associate in a ligand-independent manner (e.g., refs. 42,43). However, in the presence of the ligand the putative self-association domains on the receptors are separated by too large a distance to enable self-association within a core hexameric complex.[41] A molecular dynamics study that compared receptor self-association with the formation of receptor-ligand complexes suggested that ligand binding induces three-fold symmetry and constrains the receptor to a specific conformation that differs from that in nonligand-bound receptors.[44]

From Dimers to Clusters

Woolf and Linderman used Monte Carlo simulations to demonstrate that reversible dimerization can cause membrane proteins that normally form 1:1 complexes (either homo or heterodimers) through specific interfaces to cluster into oligomer-like structures.[45] In this view, it is important to note that dimerization is generally reversible and switching of partners to form interactions competes with diffusion to prevent interactions. When partner switching is fast relative to the diffusion rate, proteins can effectively share a single bond between multiple proteins and thus form extended clusters of proteins. This mechanism might also contribute to receptor cross talk—the simultaneous or sequential binding of multiple cell surface receptors to different ligands that results in the coordinated regulation of signal transduction across membranes.

G-Protein-Coupled Receptors

Chemokines induce chemotaxis (cell movement) by binding to G-protein-coupled receptors (GPCRs), which are the most common cell surface receptors making up ~4% of the human genome (~800 human GPCRs). These receptors are integral membrane proteins with a common core of seven transmembrane α-helices, but the extra- and intracellular domains are widely varied. There has been some debate in the field as to whether GPCRs function largely as monomers or whether oligomerization is a common requirement for biological activity (e.g., refs. 46,47). Certainly, some GPCRs can function as monomers,[48] and others appear to function as homodimers, heterodimers or higher order oligomers (e.g., refs. 49,50). At this stage many of the experiments used to support both monomer-centric and oligomer-centric views cannot be unambiguously interpreted because of the limits of the technologies and/or the use of nonnative membranes or nonphysiological concentrations of proteins.[51] However, it is reasonable that the many different GPCRs may have evolved different mechanisms of self-association and could adopt different oligomeric states that vary according to the life cycle of the molecule.[51]

Transporting Molecules across Membranes

Membrane channels, through which molecules are transported, almost invariably involve protein oligomerization. For example, aquaporins and aquaglyceroporins are integral membrane proteins that oligomerize to form homotetramers and thereby generate a central pore through which water (or water and glycerol) are selectively transported across the membrane—the channels do not transport ions (Fig. 7).[52] The association of like proteins to generate a central pore appears to be a common theme in membrane channels. Opening and closing of the pores, selectivity for target molecules and regulation of the channels are achieved by different mechanisms and include generation of selectivity filters through

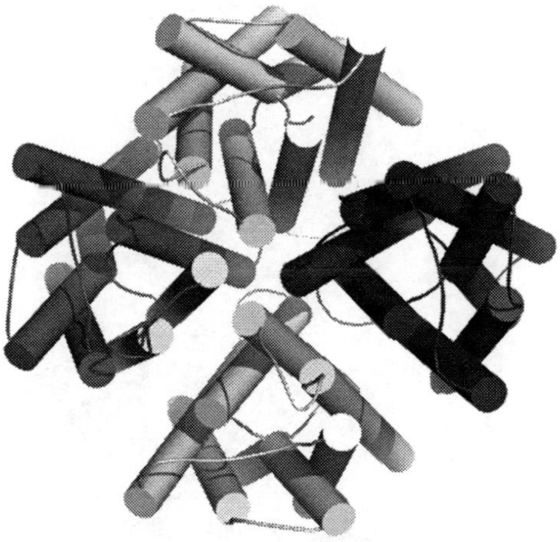

Figure 7. Human aquaporin. This integral membrane protein forms a homotetramer with a central pore through which water is transported. (PDB 3GD8).

manipulation of the common subunit interface of the multimer and binding of accessory proteins.[53] For example, potassium channels, which are an important and highly studied family of proteins that are found in all domains of life, also form tetramers (see chapter on potassium channels by Clarke and Gulbis). This theme is likely to also be true for the chloride intracellular channel (CLIC) family. CLIC proteins are a novel class of putative intracellular ion channels that are widely expressed in different intracellular compartments (see review by Singh).[54] These proteins are thought to interconvert between a soluble fold and an integral membrane form. The soluble form shares structural homology with members of the omega glutathione *S*-transferase (GST) superfamily and can undergo a redox-controlled monomer-dimer transition; the dimer is formed under strongly oxidising conditions. CLICs can auto-insert into the membrane via a single transmembrane domain to form multimeric (tetramers or higher order oligomers) ion channels with a predicted channel pore formed by the N-terminal domains.

DNA BINDING AND GENE EXPRESSION

Targeting DNA

Protein dimerization/oligomerization is a common factor in many processes that involve binding to DNA, including chromatin packaging, gene expression, DNA repair and replication (e.g., Fig. 8). In chromatin, the symmetrical histone octamer core particle consists of an H3/H4 tetramer and two H2A/H2B dimers, This construction allows for symmetrical and repeating DNA:protein contacts in the nucleosome and along the

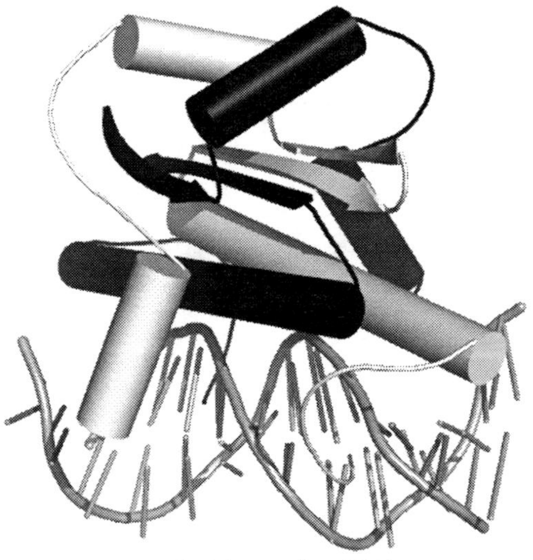

Figure 8. The transcription factor MEF2A bound to DNA. This transcription factor exists as a dimer in solution and binds to DNA as a dimer, through its MADS-box domain. Homo- and hetero-dimers formed by various members of the MEF2A family are thought to have a similar overall structure. (PDB 3KOV).

DNA.[55] Type II restriction enzymes, which are found in most bacteria and have been an essential part of recombinant DNA technology, bind to palindromic DNA sequences as homodimers; each subunit contacts one-half of the palindrome resulting in a highly symmetric protein:DNA complex. In this way protein dimerization increases the binding affinity for DNA through cooperative binding and increases binding specificity by doubling the length of the DNA recognition sequence. For a detailed discussion of how protein dimerization contributes to nucleic acid binding and transcriptional regulation see the chapters by Wilce, Vivian and Wilce and by Funnell and Crossley.

Transcription Complexes and Long Range Interactions

Dimerization can also provide physical links between different elements in the cell. For example, the transcriptional adaptor protein Ldb1 (LIM domain binding protein 1), is an important developmental protein that targets a class of transcription factors and related transcriptional regulators, the LIM-only and LIM-homeodomain proteins.[56] Ldb1 contains a self-association domain that forms trimers in solution[57] and is essential for the biological activity of the LIM proteins (e.g., ref. 58). Self-association of Ldb1 allows the formation of multi-protein complexes that can target DNA through multiple contacts. These complexes can target repeated DNA motifs with high avidity (such as multiple closely spaced homeodomain binding sites), or potentially mediate long range interactions (such as promoter-enhancer interactions). As an example of the latter case, Ldb1 was shown to promote GATA-1-mediated DNA looping at the β-globin locus control region.[59,60] The ability of the homo-oligomer to bind multiple different proteins (e.g., ref. 61) allows for similar complexes to have different properties and provides a mechanism for partner/cofactor exchange, i.e., the replacement of a protein from a complex by another to bring about a different biological activity. The self-association of Ldb1 is relatively weak (1:1 monomer to trimer at ~10^{-5} M), suggesting that Ldb1 multimers will only form at high local concentrations and/or in cooperation with other factors.[57]

CONCLUSION

Self association is a fairly general physical property of proteins that can confer many different functional advantages. Dimers and oligomers provide structure and stability, regulate activity by modulating binding sites, affinities or combinatorial specificity, facilitiate preconcentration of molecules, generate links between different cellular components and transmit signals, ions and other molecules across membranes. Given the large number of fundamental biological processes in which protein dimerization plays an essential role, there must be an enormous evolutionary pressure to develop and maintain specific dimerization interfaces in biology.

ACKNOWLEDGEMENTS

JMM is supported by a Senior Research Fellowship and MS is supported by an RD Wright Fellowship, both from the National Health and Medical Research Council of Australia.

REFERENCES

1. Marianayagam NJ, Sunde M, Matthews JM. The power of two:protein dimerization in biology. Trends Biochem Sci 2004; 29:618-625.
2. Krissinel E. Macromolecular complexes in crystals and solutions. Acta Crystallog D, Biol Crystallog 2011; 67:376-385.
3. Ispolatov I, Yuryev A, Mazo I et al. Binding properties and evolution of homodimers in protein-protein interaction networks. Nucl Acids Res 2005; 33:3629-3635.
4. Jacques D, Trewhella J. Small-angle scattering for structural biology—expanding the frontier while avoiding the pitfalls. Protein Sci 2010; 19:642-657.
5. Lukatsky DB, Shakhnovich BE, Mintseris J et al. Structural similarity enhances interaction propensity of proteins. J Mol Biol 2007; 365:1596-1606.
6. Dey S, Pal A, Chakrabarti P et al. The subunit interfaces of weakly associated homodimeric proteins. J Mol Biol 2010; 398:146-160.
7. Pearce MJ, Mintseris J, Ferreyra J et al. Ubiquitin-like protein involved in the proteasome pathway of Mycobacterium tuberculosis. Science 2008; 322:1104-1107.
8. Darwin KH. Prokaryotic ubiquitin-like protein (Pup), proteasomes and pathogenesis. Nat Rev Microbiol 2009; 7:485-491.
9. Beckett D. Regulated assembly of transcription factors and control of transcription initiation. J Mol Biol 2001; 314:335-352.
10. Pereira-Leal JB, Levy ED, Kamp C et al. Evolution of protein complexes by duplication of homomeric interactions. Genome Biol 2007; 8:R51.
11. Aragues R, Sali A, Bonet J et al. Characterization of protein hubs by inferring interacting motifs from protein interactions. PLoS Comp Biol 2007; 3:1761-1771.
12. Bahadur RP, Chakrabarti P, Rodier F et al. Dissecting subunit interfaces in homodimeric proteins. Prot Struct Funct Genetics 2003; 53:708-719.
13. Yatime L, Hein KL, Nilsson J et al. Structure of the RACK1 dimer from Saccharomyces cerevisiae. J Mol Biol 2011; 411:486-498.
14. Hopfner KP, Craig L, Moncalian G et al. The Rad50 zinc-hook is a structure joining Mre11 complexes in DNA recombination and repair. Nature 2002; 418:562-566.
15. Cahill D, Carney JP. Dimerization of the Rad50 protein is independent of the conserved hook domain. Mutagenesis 2007; 22:269-274.
16. Gilmore R, Coffey MC, Leone G et al. Co-translational trimerization of the reovirus cell attachment protein. EMBO J 1996; 15:2651-2658.
17. Christis C, Lubsen NH, Braakman I. Protein folding includes oligomerization—examples from the endoplasmic reticulum and cytosol. FEBS J 2008; 275:4700-4727.
18. Yu X, Kong Y, Dore LC et al. An erythroid chaperone that facilitates folding of alpha-globin subunits for hemoglobin synthesis. J Clin Invest 2007; 117:1856-1865.
19. Zhang X, Beuron F, Freemont PS. Machinery of protein folding and unfolding. Curr Opin Struct Biol 2002; 12:231-238.
20. Yonehara M, Minami Y, Kawata Y et al. Heat-induced chaperone activity of HSP90. J Biol Chem 1996; 271:2641-2645.
21. Jorgensen CS, Ryder LR, Steino A et al. Dimerization and oligomerization of the chaperone calreticulin. Eur J Biochem 2003; 270:4140-4148.
22. Ishikawa T, Maurizi MR, Belnap D et al. Docking of components in a bacterial complex. Nature 2000; 408:667-668.
23. Donaldson LW, Wojtyra U, Houry WA. Solution structure of the dimeric zinc binding domain of the chaperone ClpX. J Biol Chem 2003; 278:48991-48996.
24. Ogura T, Wilkinson AJ. AAA+ superfamily ATPases: common structure—diverse function. Gene Cell 2001; 6:575-597.
25. la Cour T, Kiemer L, Mølgaard A et al. Analysis and prediction of leucine-rich nuclear export signals. Protein Eng Des Sel 2004; 17:527-536.
26. Kutay U, Güttinger S. Leucine-rich nuclear-export signals: born to be weak. Trends Cell Biol 2005; 15:121-124.
27. Dong X, Biswas A, Suel KE et al. Structural basis for leucine-rich nuclear export signal recognition by CRM1. Nature 2009; 458:1136-1141.
28. Monecke T, Guttler T, Neumann P et al. Crystal structure of the nuclear export receptor CRM1 in complex with Snurportin1 and RanGTP. Science 2009; 324:1087-1091.
29. Stommel JM, Marchenko ND, Jimenez GS et al. A leucine-rich nuclear export signal in the p53 tetramerization domain: regulation of subcellular localization and p53 activity by NES masking. EMBO J 1999; 18:1660-1672.

30. Pauling L. The oxygen equilibrium of hemoglobin and its structural interpretation. Proc Natl Acad Sci USA 1935; 21:186-191.
31. Monod J, Jacob F. Teleonomic mechanisms in cellular metabolism, growth and differentiation. Cold Spring Harb Symp Quant Biol 1961; 26:389-401.
32. Cui Q, Karplus M. Allostery and cooperativity revisited. Protein Sci 2008; 17:1295-1307.
33. Wei H, Zhou MM. Dimerization of a viral SET protein endows its function. Proc Natl Acad Sci USA 2010; 107:18433-18438.
34. Brown RJ, Adams JJ, Pelekanos RA et al. Model for growth hormone receptor activation based on subunit rotation within a receptor dimer. Nat Struct Mol Biol 2005; 12:814-821.
35. Seubert N, Royer Y, Staerk J et al. Active and inactive orientations of the transmembrane and cytosolic domains of the erythropoietin receptor dimer. Mol Cell 2003; 12:1239-1250.
36. Woodcock JM, McClure BJ, Stomski FC et al. The human granulocyte-macrophage colony-stimulating factor (GM-CSF) receptor exists as a preformed receptor complex that can be activated by GM-CSF, interleukin-3, or interleukin-5. Blood 1997; 90:3005-3017.
37. Lopez AF, Hercus TR, Ekert P et al. Molecular basis of cytokine receptor activation. IUBMB Life 2010; 62:509-518.
38. Boulanger MJ, Chow D-c, Brevnova EE et al. Hexameric structure and assembly of the interleukin-6/il-6 α-receptor/gp130 complex. Science 2003; 300:2101-2104.
39. Hansen G, Hercus TR, McClure BJ et al. The structure of the GM-CSF receptor complex reveals a distinct mode of cytokine receptor activation. Cell 2008; 134:496-507.
40. Yu X, Sharma KD, Takahashi T et al. Ligand-independent dimer formation of epidermal growth factor receptor (EGFR) is a step separable from ligand-induced EGFR signaling. Mol Biol Cell 2002; 13:2547-2557.
41. Mukai Y, Nakamura T, Yoshikawa M et al. Solution of the structure of the TNF-TNFR2 complex. Sci Signal 2010; 3:ra83-.
42. Chan FK, Chun HJ, Zheng L et al. A domain in TNF receptors that mediates ligand-independent receptor assembly and signaling. Science 2000; 288:2351-2354.
43. Clancy L, Mruk K, Archer K et al. Preligand assembly domain-mediated ligand-independent association between TRAIL receptor 4 (TR4) and TR2 regulates TRAIL-induced apoptosis. Proc Natl Acad Sci USA 2005; 102:18099-18104.
44. Wassenaar TA, Quax WJ, Mark AE. The conformation of the extracellular binding domain of Death Receptor 5 in the presence and absence of the activating ligand TRAIL: a molecular dynamics study. Proteins 2008; 70:333-343.
45. Woolf PJ, Linderman JJ. Self organization of membrane proteins via dimerization. Biophys Chem 2003; 104:217-227.
46. Chabre M, Deterre P, Antonny B. The apparent cooperativity of some GPCRs does not necessarily imply dimerization. Trends Pharmacol Sci 2009; 30:182-187.
47. Milligan G. The role of dimerisation in the cellular trafficking of G-protein-coupled receptors. Curr Opin Pharmacol 2010; 10:23-29.
48. Whorton MR, Bokoch MP, Rasmussen SG et al. A monomeric G protein-coupled receptor isolated in a high-density lipoprotein particle efficiently activates its G protein. Proc Natl Acad Sci USA 2007; 104:7682-7687.
49. Maurel D, Comps-Agrar L, Brock C et al. Cell-surface protein-protein interaction analysis with time-resolved FRET and snap-tag technologies: application to GPCR oligomerization. Nat Methods 2008; 5:561-567.
50. Wilson S, Wilkinson G, Milligan G. The CXCR1 and CXCR2 receptors form constitutive homo- and heterodimers selectively and with equal apparent affinities. J Biol Chem 2005; 280:28663-28674.
51. Gurevich VV, Gurevich EV. GPCR monomers and oligomers: it takes all kinds. Trends Neurosci 2008; 31:74-81.
52. Gonen T, Walz T. The structure of aquaporins. Q Rev Biophys 2006; 39:361-396.
53. Raja M. Diverse gating in K(+) channels: Differential role of the pore-helix glutamate in stabilizing the channel pore. Biochem Biophys Res Comm 2011.
54. Singh H. Two decades with dimorphic Chloride Intracellular Channels (CLICs). FEBS Lett 2010; 584:2112-2121.
55. Cockerill PN. Structure and function of active chromatin and DNase I hypersensitive sites. FEBS Journal 2011; 278:2182-2210.
56. Matthews JM, Visvader JE. LIM-domain-binding protein 1: a multifunctional cofactor that interacts with diverse proteins. EMBO Rep 2003; 4:1132-1137.
57. Cross AJ, Jeffries CM, Trewhella J et al. LIM Domain Binding Proteins 1 and 2 Have Different Oligomeric States. J Mol Biol 2010; 399:133-144.
58. Breen JJ, Agulnick AD, Westphal H et al. Interactions between LIM domains and the LIM domain-binding protein Ldb1. J Biol Chem 1998; 273:4712-4717.

59. Song SH, Hou C, Dean A. A positive role for NLI/Ldb1 in long-range beta-globin locus control region function. Mol Cell 2007; 28:810-822.
60. Soler E, Andrieu-Soler C, de Boer E et al. The genome-wide dynamics of the binding of Ldb1 complexes during erythroid differentiation. Genes Dev 2010; 24:277-289.
61. Jurata LW, Pfaff SL, Gill GN. The nuclear LIM domain interactor NLI mediates homo- and heterodimerization of LIM domain transcription factors. J Biol Chem 1998; 273:3152-3157.

CHAPTER 2

THE DETECTION AND QUANTITATION OF PROTEIN OLIGOMERIZATION

David A. Gell,[1,2] Richard P. Grant[1] and Joel P. Mackay*,[1]

[1]School of Molecular Bioscience, University of Sydney, Sydney, New South Wales, Australia; [2]Menzies Research Institute, University of Tasmania, Hobart, Tasmania, Australia

**Corresponding Author: Joel P. Mackay—Email: j.mackay@mmb.usyd.edu.au*

Abstract: There are many different techniques available to biologists and biochemists that can be used to detect and characterize the self-association of proteins. Each technique has strengths and weaknesses and it is often useful to combine several approaches to maximize the former and minimize the latter. Here we review a range of methodologies that identify protein self-association and/or allow the stoichiometry and affinity of the interaction to be determined, placing an emphasis on what type of information can be obtained and outlining the advantages and disadvantages involved. In general, in vitro biophysical techniques, such as size exclusion chromatography, analytical ultracentrifugation, scattering techniques, NMR spectroscopy, isothermal titration calorimetry, fluorescence anisotropy and mass spectrometry, provide information on stoichiometry and/or binding affinities. Other approaches such as cross-linking, fluorescence methods (e.g., fluorescence correlation spectroscopy, FCS; Förster resonance energy transfer, FRET; fluorescence recovery after photobleaching, FRAP; and proximity imaging, PRIM) and complementation approaches (e.g., yeast two hybrid assays and bimolecular fluorescence complementation, BiFC) can be used to detect protein self-association in a cellular context.

INTRODUCTION

Protein self-association covers a spectrum of behaviours: specific, nonspecific, reversible and irreversible. Specific interactions have a defined subunit arrangement and stoichiometry and are generally mediated by interfaces that display large numbers of noncovalent interactions. In contrast, nonspecific interactions often give rise to amorphous aggregates with no defined stoichiometry. The purpose of this chapter is to outline a

Protein Dimerization and Oligomerization in Biology, edited by Jacqueline M. Matthews.

number of experimental methods that are available to identify protein self-association and determine the stoichiometry and affinity of the interaction. The approaches range from structural and biophysical through to biochemical and cellular in nature. Cellular methods have the advantage of being carried out in a more physiological environment, but knowledge of all influencing factors is generally incomplete. In contrast, in vitro methods allow a quantitative and unambiguous analysis in highly purified solutions, but require some extrapolation to draw inferences about biological relevance. In general, no technique is suited to all situations and it is good practice to combine measurements to take advantages of their respective strengths.

BIOPHYSICAL METHODS

General Considerations

Before detailing the methods, a number of general points can be made by considering a typical reversible homodimerization reaction:

$$A + A \Leftrightarrow B \tag{1}$$

The equilibrium dissociation constant for the reaction is a measure of the thermodynamic stability of the complex and is defined as

$$K_D = \frac{[A]^2}{[B]} \tag{2}$$

where $[A]$ and $[B]$ are the molar concentrations of the monomer and dimer, respectively. Equation 2 reveals that as the total concentration of the solute increases, $[B]$ must increase faster than $[A]$. For a particular system of interest, an equilibrium reaction can be confirmed and a value of K_D obtained by measuring some property of the solution that is sensitive to the ratio $[A]/[B]$ over a range of solute concentrations. One such property is average molecular weight. In fact, there are a number of ways to calculate an average molecular weight depending on the physical property under investigation. For example sedimentation, classical light scattering and X-ray scattering (see sections “Analytical Ultracentrifugation” and “Scattering”) all naturally yield the weight-average molecular weight, which is defined as

$$M_w = \frac{\sum_i c_i M_i}{\sum_i c_i} = \frac{\sum_i c_i M_i}{c_{tot}} \tag{3}$$

where c_i is the weight concentration of the species with molecular weight M_i and c_{tot} is the total concentration of solute.

An important practical point arises here. At a total monomer concentration 10, 10^2, or 10^3 times K_D, the molar fraction [free monomer]/[total monomer] is 0.2, ~0.07 and 0.02 respectively. Thus, in order to observe significant shifts in the equilibrium, measurements must be made at concentrations in the order of K_D. For very stable complexes, this requirement can be beyond the sensitivity of many physical methods. At concentrations well above the K_D, M_w corresponds closely to that of the pure oligomer and can therefore be used to determine the subunit stoichiometry.

Other physical properties sensitive to the oligomerisation reaction are hydrodynamic size, sedimentation coefficient and translational and rotational diffusion coefficients and measurement of these is discussed in the following sections.

SIZE EXCLUSION CHROMATOGRAPHY (SEC)

An SEC column contains hydrated beads made from cross-linked polymers such as dextran, agarose or polyacrylamide. The migration of a protein through the column is governed by its partitioning between two solvent spaces, the solvent spaces surrounding the packed beads and the solvent within the porous beads. For any given species the volume fraction of the internal bead volume that is accessible is termed the partition coefficient σ_p (see ref. 1). Smaller proteins have a larger value of σ_p and require a greater volume of buffer to pass through the column before they are eluted.

Empirically, σ_p correlates best with the Stokes radius (R_H; the radius of the hard sphere with the same translational diffusion coefficient) of the species.[2] For monodisperse solutions of spherical globular proteins molecular weight can be predicted from R_H with reasonable accuracy.[3] In this case σ_p correlates well with molecular weight according to

$$\sigma_p = -A \log M + B \tag{4}$$

where A and B are column-specific constants that can be determined from a plot of σ_p against $\log M$ for a range of well-characterized protein standards.[4] The molecular weight for the protein of interest can then be determined from this standard curve. Unfortunately, for nonspherical species σ_p is a poor predictor of M. However, SEC can provide M and molecular shape information by combination with, for example, sedimentation coefficient[2] or in-line light scattering measurements.

For interacting systems the appearance of an SEC elution profile depends on the rate of exchange between monomer and oligomer states. In general, self-association reactions with $K_D < 1$ μM or $K_D > 1$ mM will result in single peaks corresponding to the oligomer or monomer respectively.[5] In contrast, migration of complexes with intermediate K_D be governed by properties of both monomer and oligomer and the elution volume will be a function of the loading concentration (Fig. 1), the K_D and the length and partitioning properties of the column. A computer simulation of SEC behaviour has been developed that models the transport of monomer and oligomer states and the continual shift in the equilibrium distribution due to diffusion and dispersion.[5] This method has been used to obtain K_Ds in the mM-μM range for dimerization of antibody light chains and antibody:antigen interactions.[6]

A modification of the SEC method, termed large zone SEC, was also developed to study weak homo-interactions of haemoglobin subunits.[1] A large sample volume (~30-60% of the total column volume) is applied such that a plateau concentration equal to the load concentration migrates through the column. Because the plateau region of the peak is not diluted the elution volume is determined by $\bar{\sigma}_p$ (the weight average of σ_p for the interacting species) and truly reflects M_w at the load concentration. Plotting $\bar{\sigma}_p$ as a function of the load concentration produces a dissociation curve from which K_D can be extracted. The method recently been employed to investigate monomer-dimer-tetramer equilibria of a

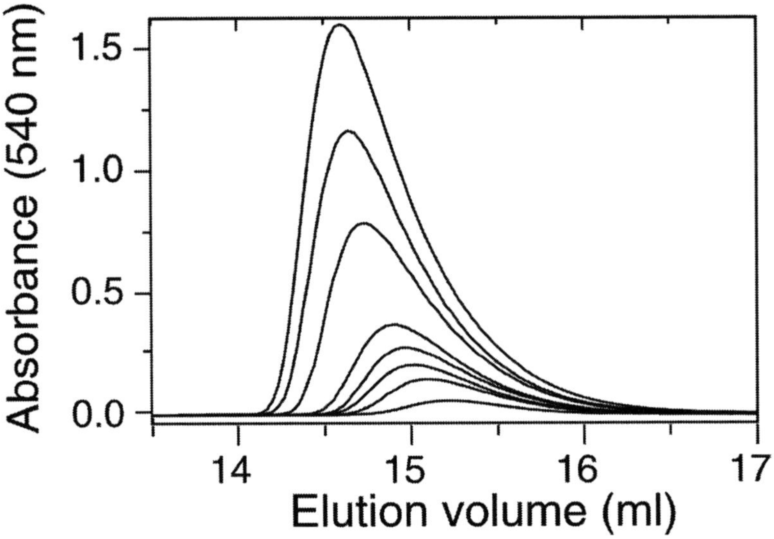

Figure 1. SEC of purified human α-haemoglobin. The elution volume of the αHb peak decreases with increasing load concentration (0.25-5.0 mg ml^{-1}), indicative of self-association.

histone-like protein H-NS,[7] dimerization of cro repressor from bacteriophage lambda (K_D = 300 nM)[8] and dimerization of the capsid protein CA from HIV-1 (K_D = 10 μM).[9]

ANALYTICAL ULTRACENTRIFUGATION (AUC)

Introduction to Sedimentation

A simple mechanical description of forces experienced by a solute particle in a centrifugal field can be applied to arrive at the following equation defining the sedimentation coefficient[10]

$$s = \frac{u}{\omega^2 r} = \frac{M(1 - \bar{v}\rho)}{N_A f} \tag{5}$$

where u is the radial velocity of solute particles, $\omega^2 r$ is the centrifugal field strength, f is the frictional coefficient, $\bar{v}$ is the partial specific volume of the particle (the inverse of its density) and ρ is the density of the solution. Values for $\bar{v}$ and ρ are usually calculated on the basis of chemical composition from physical data tables.[11] The unit of s is the Svedberg (1 S ≈ 10^{-13} s).

Diffusion acts to oppose redistribution of solute due to sedimentation. However, if the centrifugal field is strong enough there will be a net migration of particles to the bottom of the tube and a moving boundary is set up between the solution and a zone of cleared solvent behind. The AUC instrument allows this moving boundary to be observed in real time (Fig. 2A,B), via absorption, interference or fluorescence optics, and values for s and the diffusion coefficient, D, to be obtained.

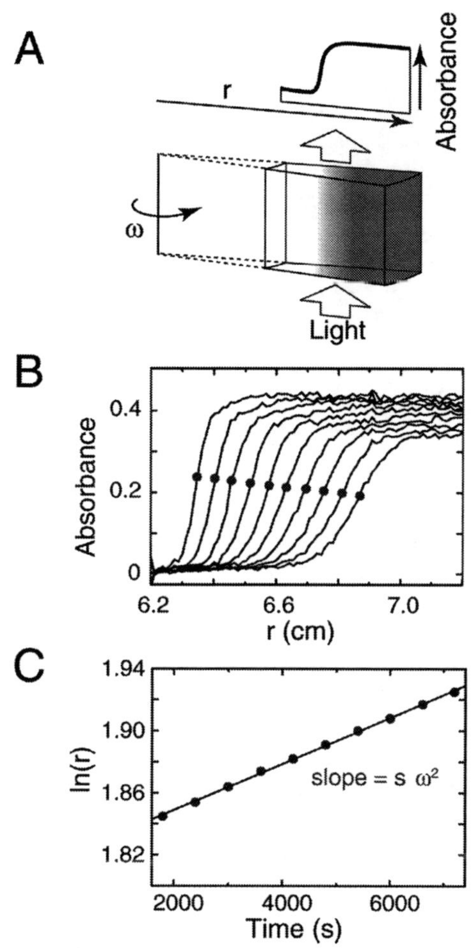

Figure 2. Sedimentation velocity. A) Schematic representation of the AUC cell, rotating at angular velocity ω. The cell contains transparent windows through which a light beam can pass during operation of the rotor. A concentration gradient of solute forms (grey shading) and can be detected by optical absorbance measurements made at incremental values of *r*. B) Successive absorbance scans (from left to right) recorded at fixed time intervals. The position of the second moment of each boundary (indicated by the dots) is used to calculate the sedimentation coefficient, *s* as shown in C.

Sedimentation Velocity

The sedimentation coefficient can be determined from the migration of the central point, or second moment, of the sedimentation boundary according to

$$\ln\frac{r}{r_0} = \omega^2 st \qquad (6)$$

where r and r_0 are the positions of the boundary at time t and $t = 0$ (the meniscus), respectively. A plot of $\ln r$ against t will produce a straight line with slope $\omega^2 s$ (Fig. 2C). For mixtures or interacting systems this method yields the weight average s. Because the boundary may be broadened by both diffusion and sample heterogeneity the term apparent sedimentation coefficient is used. For a single ideal solute, a decrease in *s* with increasing concentration is expected on theoretical grounds.[12] An increase in *s* with concentration is a firm indication of protein self-association.

A number of sophisticated mathematical tools have been developed to analyze sedimentation velocity data in greater detail.[13,14] These methods generally yield weight averaged values for *s* and hence experiments performed at different plateau concentrations allow thermodynamic parameters to be obtained for complexes in fast exchange; examples include dimerization of cytomegalovirus protease ($K_D = 17\ \mu M$)[15] and trimer-hexamer-12mer association reactions of the chaperone gp57A.[16] Sedimentation velocity has been used to determine the kinetics for competing heterodimerization reactions[17] and even multiple self-association mechanisms with distinct kinetics for a single protein.[18]

In combination with direct measurements of *M* by sedimentation equilibrium (see below), *s* values yield *f* (Equation 5), which can provide some information about molecular shape.[19] This approach has been used to propose hexameric ring structures for *E. coli* DNA replication protein (RepA)[20] and *N*-ethylmaleimide-sensitive fusion protein (NSF).[21]

Sedimentation Equilibrium

This method can be used to obtain a shape-independent molecular weight for a macromolecule and is often the most suitable method for obtaining stoichiometries and equilibrium constants for rapidly reversible self-associating systems with dissociation constants in the range 100 nM-10 mM (see ref. 22). Sedimentation equilibrium data are frequently used to support other molecular weight measurements or to facilitate interpretation of shape-sensitive data obtained from techniques such as sedimentation velocity (see for example refs. 15, 16, 18, 20, 21, 23).

At rotor velocities sufficiently slow that the protein does not simply pellet at the bottom of the cell, equilibrium is reached between sedimentation and diffusion (Fig. 3). Because there is no net flux of material, terms relating to the frictional coefficient *f* disappear (see Equation 5) and the radial concentration gradient, $c(r)$, of a single monomeric species becomes a simple function of the monomer molecular weight according to[3,10]

$$c(r) = c(r_0)\exp\{\psi M(r^2 - r_0^2)\} \tag{7a}$$

where

$$\psi = \omega^2(1 - \bar{v}\rho)/2RT \tag{7b}$$

and $c(r_0)$ is the concentration at the reference point (usually the meniscus), R is the gas constant and T is the absolute temperature. Thus, the most basic analysis of sedimentation equilibrium data is a plot of ln*c* versus r^2. For a single species the plot is a straight line, with a gradient proportional to the molecular weight. For mixtures of solutes the final concentration profile will be a sum of exponentials and the ln*c* versus r^2 plot will deviate from a straight line; the gradient at r is then proportional to M_w at that point in the cell.

Obtaining an equilibrium constant requires global fitting of data obtained across a range of sample loading concentrations and rotor speeds to a specific interaction model using non linear least squares algorithms.[24] For a monomer-n-mer interaction

$$c_{tot}(r) = c(r_0)\exp\{\psi M(r^2 - r_0^2)\} + K_D^{-1}[c(r_0)]^n \exp\{\psi n M(r^2 - r_0^2)\} \quad (8)$$

where c_{tot} is the total weight concentration of all reactants. Equation 8 can be solved for K_D if the monomer weight is known. Expressions similar to Equation 8 can be generated for other, more complex, interaction models.[22,24] The distribution of the residuals between observation and fit at each position across the cells is an important indicator of the suitability of a given model (Fig. 3B).

SCATTERING—RAYLEIGH SCATTERING OF VISIBLE LIGHT

Rayleigh theory describes light scattering from particles much smaller than the wavelength of the incident light. The incident electromagnetic wave induces an oscillating dipole in the particle, the magnitude of which depends on the electrical polarizability of the particle. The oscillating dipole then emits radiation with the same frequency as the incident beam (elastic scattering). Rayleigh scattering of plane-polarised light is depicted in Figure 4—note that the scattering intensity *I*, is independent of the scattering angle

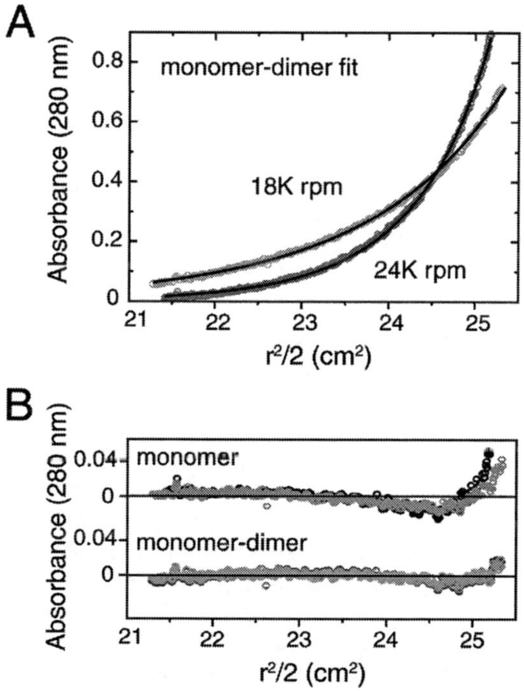

Figure 3. Sedimentation equilibrium. Analysis of purified α-haemoglobin performed at 4°C in 150 mM NaCl, 20 mM sodium phosphate, pH 7.0. A) Data were recorded at 18,000 and 24,000 rpm and a global fit to various interaction models was performed using a nonlinear least squares algorithm implemented by the software NONLIN.[99] The fit to a monomer-dimer model is shown (black lines). B) Residuals for fits to monomer and monomer-dimer models. A monomer model results in large systematic discrepancies between fit and the data (18,000 rpm, light circles; 24,000 rpm, dark circles). A monomer-dimer model (K_D = 0.5 mM) gives an improved fit and is consistent with measurements by large zone SEC.[1]

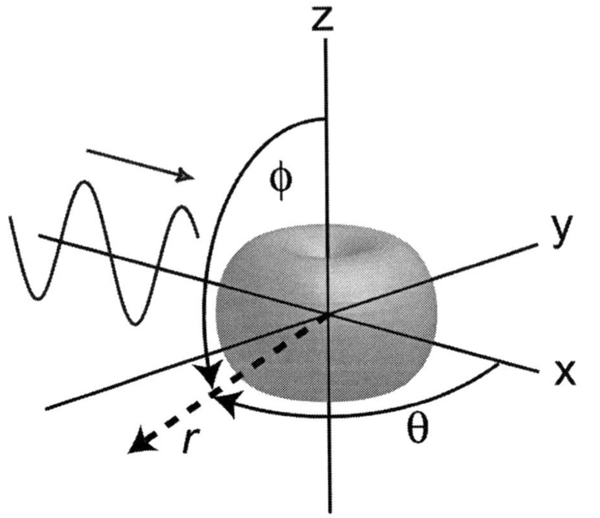

Figure 4. Light scattering from a small particle located at the origin with dimensions much less than the incident wavelength. If the incident light wave is plane polarized along the z-axis, the scattering intensity is dependent upon the angle ϕ and the sample-to-detector distance, r. The distance from the origin to the shaded surface is proportional to the light scattering intensity in that direction, with an intensity of zero directly along the z-axis. Scattering intensity is independent of the angle θ, measured in the xy plane.

θ, measured perpendicular to the plane of polarisation. To remove the influence of the instrumental parameters, incident light intensity I_O and sample-to-detector distance r, scattering intensity is expressed as a Rayleigh ration $R(\theta)$, defined

$$R(\theta) = \frac{I_\theta r^2}{I_0} \tag{9}$$

For visible light scattering (LS) we generally consider particles to be "small" if the root mean square radius, R_G, is less than ~15 nm (R_G is a measure of molecular size describing the distribution of mass). For such particles

$$R(\theta) = K_{LS} c M_w \tag{10}$$

where c is the particle concentration and K_{LS} is a constant for protein solutions. It includes a factor relating to the increased electrical polarizability of the solution due to the presence of the macromolecule. For proteins, polarizability depends only upon molecular size and is essentially independent of the amino acid sequence.[25] Hence, M_w can be determined directly from $R(\theta)$ as long as c can be measured accurately (usually on the basis of refractive index; RI). As $R(\theta)$ is the sum of the Rayleigh ratios for all particles present in the solution, LS measurements yield the weight-average molecular weight, M_w.[10]

Equation 10 is based on two assumptions: (1) that scattering from each particle is unaffected by its neighbours and (2) that the particles are small ($R_G < \lambda/20$). Assumption 1 is generally a good approximation for dilute protein solutions (< 0.1 mg/ml at the detector[25]). Scattering intensity can be measured at a range of concentrations and extrapolated to zero concentration to detect non-ideal behaviour. For larger particles Rayleigh-Debye-Gans

(RGD) theory predicts that scattering at zero angle $R(0)$ should be identical to that for a small particle with the same mass.[26] Scattering at zero angle cannot be measured due to the intensity of the incident beam, but $R(\theta)$ determined at multiple angles (multiangle laser light scattering; MALLS) can be extrapolated back to $R(0)$ to determine M_w.[10]

Placing LS and RI/UV detectors in-line with a size exclusion chromatography system allows the M_w for eluting peaks to be determined rigorously in a way which is not possible on the basis of elution volume, making this a powerful system for studying protein complexes (Fig. 5). For protein complexes at concentrations well above their K_D, the M_w can be used to infer the subunit stoichiometry.[20,21] For weaker complexes the M_w of an eluting peak may be intermediate between the free and bound proteins. For

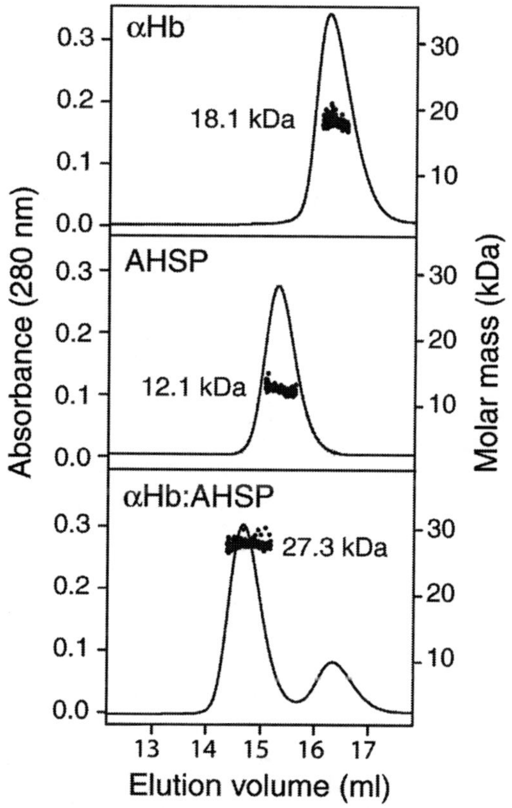

Figure 5. Determining M_w for protein complexes by SEC-MALLS. Data are shown for α-haemoglobin (αHb), αHb-stabilizing protein (AHSP) and the αHb:AHSP complex (1.2:1 molar ratio). Calculations of M_w for 0.05 ml fractions across the elution peaks (black dots) were based on LS measurements at three angles (mini-DAWN with 690 nm laser, Wyatt Technology Corp., Santa Barbara CA) and concentrations from refractive index (Optilab differential refractometer, Wyatt). Numbers printed next to the main peaks are the average M_w values (kDa) for the whole peak. The value of M_w for αHb is significantly above the predicted value of 15.3 kDa, consistent with the presence of weak self-association (see Fig. 1). The M_w for AHSP is close to the predicted value of 12.0 kDa, indicating a monomer. The early elution time for AHSP is indicative of a nonspherical shape. The M_w of the αHb:AHSP complex corresponds closely to the predicted value for a heterodimer (27.2 kDa; the small free αHb peaks is consistent with the molar mixing ratio).

example, erythropoietin (EPO) forms complexes with either one or two EPO receptors with different affinities.[27] LS can also be used to determine the stoichiometry of protein complexes that form in detergent micelles.[28]

DYNAMIC LIGHT SCATTERING (DLS)

The measurement of LS described in the previous section assumes a time averaged intensity measurement. However, on the microscopic scale, local variations in solute concentration will occur due to Brownian motion. If a small volume of sample is illuminated, these concentration changes are measurable as fluctuations in scattering intensity. The rate of fluctuation can be related simply to the translational diffusion coefficient.[10] The diffusion coefficient is a function of the size and the shape of the particle and can be expressed in terms of the Stokes radius, R_H and converted to a molecular volume for simple geometric shape such as a sphere.[3,10] Notably, R_H relates to the hydrated protein and hence the corresponding molecular volume will include associated water molecules. Assumptions about molecular shape and hydration therefore have to be made in order to calculate molecular weight,[19] making DLS an unreliable method for this purpose. One strength of the DLS method is the ability to simultaneously characterize particles over a huge size range, but it cannot generally resolve species that display less than a two-fold difference in diffusion coefficients,[29] corresponding to a ~8-fold difference in molecular weight.

Although DLS has limited resolving power for mixtures, it is a rapid and sensitive method to detect changes in R_H for a single species—for example conformational changes in calmodulin upon binding to target peptides.[30] DLS is well suited to studying protein aggregation, in relation to protein function (e.g., B_L-crystallin aggregation[31]) or sample preparation (e.g., to assess polydispersity of protein samples prior to crystallisation trials[32]). Due to the ease, rapidity and minimal sample requirements DLS is also often used to demonstrate that oligomeric structures observed in the crystals are also present in solution.

SMALL ANGLE X-RAY SCATTERING (SAXS)

SAXS is a powerful technique for studying protein size and shape in solution.[33] As with visible light, elastic scattering of X-rays can be described using classical Rayleigh theory.[34] X-rays interact with electrons causing them to oscillate and emit radiation with the same frequency as the illuminating beam. The difference in electron density between the solute and the solvent ($\Delta\rho$) and the volume of the particles are therefore the critical parameters determining X-ray scattering intensity. As with LS, M_w and R_G can be obtained directly from SAXS data by extrapolation of scattering to zero angle.[35] The resolving power of scattering techniques is determined by the ratio R_G/λ, which imposes a lower limit >15 nm for determination of R_G by LS ($\lambda \approx 690$ nm). By comparison, R_G for even the smallest protein can be accurately measured at X-ray wavelengths ($\lambda \approx 0.1$ nm).

Although SAXS data can be modeled to obtain 3-D structure information,[33] model free measurements of M_w and R_G provide insights into self-association mechanisms. For example, Rescic et al[36] found that R_G and R_H determined by SAXS and DLS, respectively, had similar values in the monomeric and the dimeric forms of human serum albumin, suggesting that asymmetric monomers align along their long axes. The transcription factor sap1 was found to be an elongated dimer of 46 kDa, with R_G consistent with a predicted coiled-coil structure.[37] Whitten et al identified the histidine kinase KinA and its negative regulator Sda as dimers by SAXS, but additionally found by using contrast variation neutron scattering methods that the SdaA binds as two separated monomers to KinA.[38] Finally, SAXS can be used effectively to study very large oligomers[39] or protein oligomers within detergent micelles.[40]

ISOTHERMAL TITRATION CALORIMETRY (ITC)

The uptake or evolution of heat that occurs when proteins associate or dissociate can be detected by highly sensitive calorimeters. A schematic diagram of an isothermal titration calorimeter is shown in Figure 6A. A standard instrument consists of two thermally isolated chambers. The reference cell contains simply water or buffer (left-hand chamber in Fig. 6A). For the analysis of the interaction between two molecules (which can be called receptor and ligand), a solution of the receptor is placed in the sample cell and a more concentrated solution of the ligand is placed in a syringe. The stepwise addition of the second molecule into the first results in heat being given out or taken up. During the entire process, a small current (< 1 mW) is applied to both the reference and the sample cell and the instrument measures the amount by which the current applied to the sample cell must be altered to maintain the same temperature as the reference cell. The change in current is a direct measure of the heat taken up or given out by the formation of the complex and this in turn

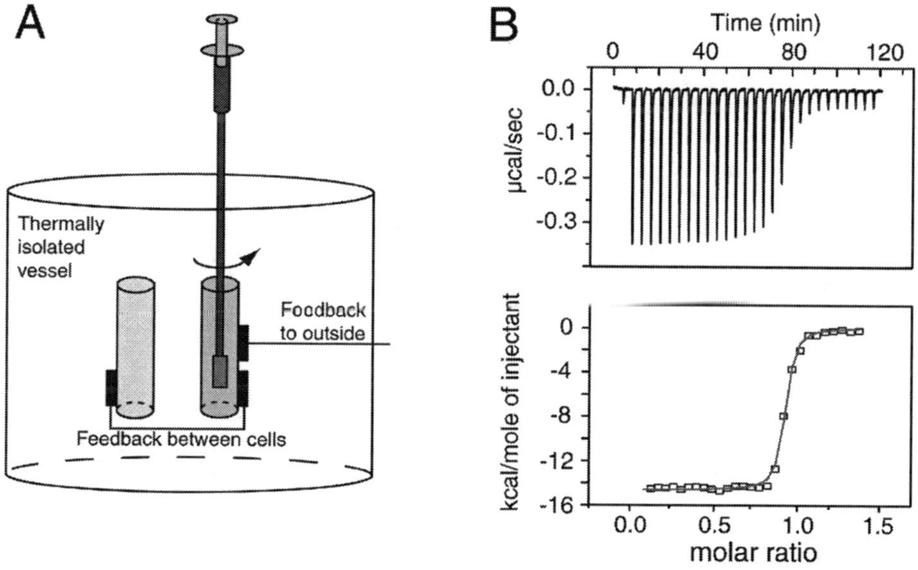

Figure 6. Isothermal titration calorimetry. A) Schematic diagram of an isothermal titration calorimeter. A syringe containing a solution of one molecule is injected in a stepwise fashion into a cell containing a binding partner. The syringe also incorporates a paddle that acts as a stirrer. B) Example of typical ITC data for a protein-DNA complex. The top trace shows heat given out from the sample cell as a function of time. Each spike corresponds to an addition of protein to the DNA.

is directly proportional to the net change in the amount of the ligand that has bound to the receptor. A plot of evolved heat as a function of the concentration of added ligand yields a binding isotherm that can be fitted to various models of binding by nonlinear least squares fitting (Fig. 6B), to derive an association constant for the interaction.

ITC has been used extensively to quantitate interactions between two different molecules (for a recent review, see ref. 41). However, it is also feasible under some circumstances to use ITC to measure self-association processes. An experiment of this type can be carried out by using a concentrated stock solution of the protein of interest in the injection syringe and titrating it into a matched buffer. The consequent dilution of the protein will shift the monomer-oligomer equilibrium towards the monomer and the calorimeter will measure any heat that is either evolved or taken up. The degree of dilution will decrease as the titration proceeds. For example, the dimerization constant of interleukin-8 measured by this method (18 μM) was in good agreement with a value obtained using analytical ultracentrifugation.[42] Similar experiments have been carried out to measure the self-association of Hexim1 (K_D = 790 μM)[43] and several vancomycin-class glycopeptide antibiotics (K_Ds in the range 5 mM-20 nM).[44] In general, ITC will most likely only be useful for a reasonably limited range of self-association constants (K_D ~0.1 mM-1 μM)—concentrations less than ~10 μM generally give signals that are very difficult to detect by ITC.

NMR SPECTROSCOPY

Introduction

It is possible to use NMR spectroscopy to provide information on protein self-association. The most detailed approach involves full structure determination, which requires a detailed interpretation of nuclear Overhauser effects (NOEs) and the ability to distinguish between intra- and inter-subunit NOEs. Other commonly used NMR based methods involve the measurement of translational[45] or rotational[46] diffusion coefficients. These approaches provide rapid estimates of hydrodynamic size, but their accuracy is limited by the dependence of R_H both on molecular weight and on molecular shape and hydration (As noted in the section "Dynamic Light Scattering"). Here we describe the use of asymmetric labeling strategies and calculation protocols that incorporate ambiguous NOE constraints, which has become widespread for the structural analysis of oligomeric proteins.

Nuclear Overhauser Effect Data for Determination of Multimeric Structures

The determination of protein structures by NMR spectroscopy is complicated significantly if the protein forms a symmetrical homodimer or higher order multimer, because of difficulties with the unambiguous interpretation of the NOE data that are routinely used to determine protein structures. The NOE is an interaction between two nuclei, usually two hydrogen atoms, that arises if the nuclei are sufficiently close in space (usually less than ~6 Å). Hundreds or thousands of such interactions can be measured for a typical well-ordered protein and they comprise the primary form of experimental data used to determine protein structures by NMR spectroscopy.

In the case of a symmetrical homodimer, it is often difficult or impossible to know whether a given NOE reports on a connection between two protons from the same protomer or between protons from two different protomers, especially in the case of

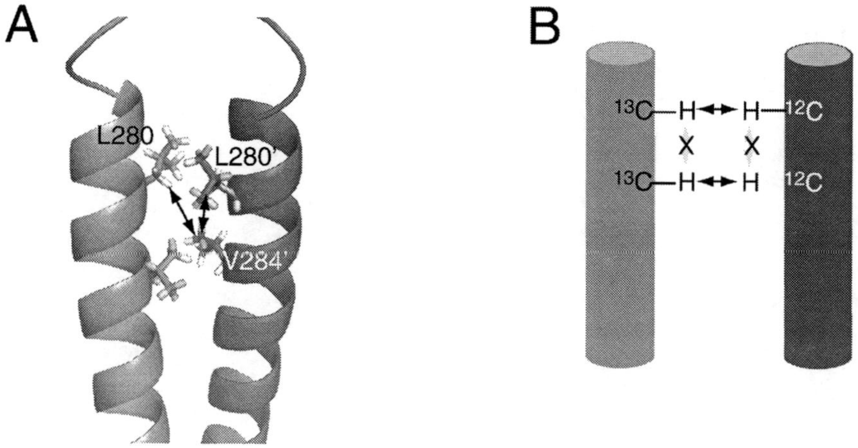

Figure 7. Detection of intermolecular NOEs in homodimers. A) Portion of the symmetric coiled coil of cJun (PDB: 1jun)[100] illustrating the difficulty in distinguishing between intramolecular and intermolecular NOEs. The sidechain protons of V284' will give NOEs to L280 in both the same monomer and in the other monomer (arrows). B) Example of a labeling scheme for the detection of intermolecular NOEs.

structures such as parallel coiled-coils, in which the same parts of each protomer are in contact with each other (Fig. 7A).

Two distinct strategies have been developed to deal with this problem. A computational approach was developed by Nilges,[47] in which all NOEs were considered to be ambiguous during the structure calculation procedure—that is, they could arise from either *intra*monomer or *inter*monomer interactions. This approach can be extended to higher order oligomers and successful applications have been demonstrated for the tetrameric and hexameric structures of the p53 tetramerization domain[48] and insulin,[49] respectively.

The second approach revolves around the preparation of protein samples with mixed isotopic labeling schemes.[50] Experiments can be used to selectively detect only NOEs between pairs of protons in which one of the pair is from a labelled protomer and the other is not[51] (Fig. 7B). This strategy has been used numerous times for the determination of both heterodimeric and homodimeric structures (for example, see ref. 52) and even in a protein trimer.[53]

MASS SPECTROMETRY

Over the preceding decades, the utility of mass spectrometry to protein science has increased dramatically. This has coincided with the development of the relatively soft ionization techniques electrospray and matrix assisted laser desorption ionization (MALDI) that have allowed large proteins to 'fly' in the mass spectrometer. In electrospray ionization sources, a solution of the protein of interest is sprayed through a very fine needle that is held at a high voltage. The drops are evaporated and then introduced into the mass spectrometer and it is possible to carry out the entire process under relatively gentle conditions (neutral pH, no organic solvents). Thus, electrospray (and the miniaturized 'nanospray' version in particular) can be used to ionize intact noncovalent macromolecular complexes and to therefore determine their stoichiometry. For example, an early study[54] showed that the tetrameric forms of avidin, concanavalin A and hemoglobin were all able to be observed directly and that the molecular masses determined were within ~0.1% of the theoretical values. The ability to measure masses with such high accuracy has permitted detailed analyses of the folding and stability of proteins such as transthyretin,[55] for which dissociation of the tetramer can lead to the formation of amyloid fibrils. Stoichiometries of very large complexes such as the bacterial chaperone GroEL (which forms an 800-kDa 14-mer) have been established using this approach,[56] making it widely applicable.

Mass spectrometry is clearly the most accurate and precise method for determining solution molecular weight, but currently the acquisition of mass spectra of large noncovalent complexes is technically challenging and perhaps therefore not as widely available as some other techniques. Nevertheless, a protocol for the analysis of noncovalent protein complexes, optimized for the use of commercial mass spectrometers has recently been reported,[57] suggesting that these experiments are within reach of an increasing number of researchers.

FLUORESCENCE ANISOTROPY

Similar to isothermal titration calorimetry, fluorescence anisotropy is most commonly used as a method for measuring association constants for heterotypic interactions, but it is also suited to studies of homodimerization.[58] When a solution containing a fluorophore is excited with polarized light, only the fluorophores that have their transition dipoles parallel to the plane of polarization will be excited (a process known as photoselection). When these molecules subsequently emit fluorescent photons (a few ns later), the degree of polarization of the emitted light will depend on the amount of rotation undergone by the molecules in the interim. Small molecules will rotate faster and therefore the emitted light will be significantly depolarized. Dimerization or the binding of a partner molecule will reduce the rate of rotational motion and the emitted light will retain a higher degree of polarization. Polarization (P) and anisotropy (r) are simply two different measures of this phenomenon, defined as:

$$r = \frac{I_{\parallel} - I_{\perp}}{I_{\parallel} + 2I_{\perp}} \tag{11a}$$

and

$$P = \frac{I_{\parallel} - I_{\perp}}{I_{\parallel} + I_{\perp}} \tag{11b}$$

where $I_{\parallel}$ and $I^{\perp}$ are the intensities of the emitted light parallel and perpendicular to the excitation source, respectively. The slower the rotation of the fluorescent species, the higher the value of r (which has a maximum value of 0.4).

Although the intrinsic fluorescence of a protein could conceivably be used for these measurements, fluorescent tags (through a unique cysteine if possible) such as fluorescein or derivatives result in a better signal-to-noise ratio. In order to measure a dimerization constant, the fluorescence anisotropy of a labeled protein is measured as a function of protein concentration. This approach has been used to measure the dimerization of a range of proteins including the transcriptional regulators MyoD,[59] TAL1[60] and NFAT.[61] These proteins exhibit dissociation constants of between 1 μM and 1 nM and both higher and lower dimerization constants should be accessible in principle, although the addition of a fluorophore (which is generally hydrophobic) can often reduce the solubility of the protein to which it is attached.

IN VIVO METHODS

Introduction

The methods described above all are all carried out using highly purified proteins in an in vitro environment. However, because experimental conditions in such cases will differ from the conditions found in the cell and extracellular milieu, it is clearly also important to establish the self-association properties of a protein in its physiologically relevant environment. This section describes a number of methods that are available to examine self-association in a cellular context.

CHEMICAL CROSS-LINKING

Chemical cross-linking has long been used for the analysis of protein self-association. At the crudest level it is possible to cross-link all instantaneously interacting species in a cell using a broad-spectrum reagent such as formaldehyde. Proteins of interest can then be separated from the complex cellular mixture by immunoprecipitation or, if exogenous constructs are expressed in the cell, using any of a number of affinity methods (GST, Ni-NTA, FLAG etc.). The sizes of cross-linked complexes are then assessed by SDS-PAGE/Western blotting. For example, the treatment of cultured cells with formaldehyde and retrieval of complexes by immunoprecipitation revealed that ATM kinase is a homodimer that dissociates upon radiation-induced phosphorylation.[62]

However, all interacting proteins in the cell are likely to be cross-linked by this method, which will not distinguish a genuine homodimerization event from some other interaction. A refinement of this technique involves specific cross-linking of cysteine residues using (for example) maleimides, methanethiosulfonates or the oxidative cross-linking reagent copper(II) *o*-phenanthroline.[63] Cysteine-free and specific cysteine-containing variants were cross-linked in vivo using a combination of these reagents to probe the dimerization interface of the *E. coli* membrane transporter ProP. The authors were able to show that ProP forms an antiparallel coiled-coil and participates in a monomer:dimer equilibrium in vivo.

FLUORESCENT METHODS

The ready availability of endogenously fluorescent proteins (such as variants of the green fluorescent protein—GFP—from *Aequorea victoria* jellyfish) provides a number of elegant and powerful methods for assessing self-association in vivo; these methods can also provide access to real-time dynamics.[64]

It is essential to ensure that the GFP-tagged protein retains its native activity and that expression levels do not cause undesirable effects; this should be determined empirically for each protein, but as an example GFP-tagged histone H2B at 10% of the total H2B protein gives bright fluorescence without detectable adverse effects on cellular function.[65] The tagged proteins should also have had sufficient time to fully equilibrate with the native pool.

Although none of the following techniques is trivial to perform, the information obtainable is very valuable in assessing the importance of the self-association of a protein in vivo.

Fluorescence Correlation Spectroscopy (FCS)

FCS relies on a conventional confocal-type microscope, but rather than scanning a relatively large field of view containing several cells, a small, well defined area (~0.2 femtolitres, equivalent to a single mitochondrion) is scanned at high temporal resolution. The observed signal arises from fluctuations in the fluorescence intensity, which correspond to individual fluorescent molecules entering and leaving the detection volume. A diffusion coefficient can therefore be determined as described for DLS (see Section "Dynamic Light Scattering").[66] Using this technique, Philip et al demonstrated that the Type 2 bradykinin receptor exists as dimers or higher-order oligomers in stimulated HEK293 cells.[67]

A limitation of the method is the inability to distinguish homo- from hetero-interactions. To do this, an alternative treatment of the fluorescence intensity trace can be employed. In fluorescent-intensity distribution analysis (FIDA), the total photon count over a period much shorter than the diffusion time is measured repeatedly to obtain a photon count histogram (PCH; number of counts versus probability). The PCH can be transformed into a plot of particle brightness versus particle number. For example, a monomer-dimer system will give rise to one peak corresponding to monomer and a second peak corresponding to dimer particles with twice the brightness.[68] Saffarian et al used this approach to show that oligomerization of EGF receptors in living cell membranes depends on cholesterol load.[69] Note that the presence of native, unlabelled protein will give an artificially low value for the number of dimers (an equal number of labelled and unlabelled subunits will result in only 25% of homodimers containing two fluorescent molecules). Thus, either the mathematical model must take this into account or the experiment must be performed in cells lacking native protein.

Förster Resonance Energy Transfer (FRET)

If the emission energy of an excited fluorophore overlaps with the excitation peak of a second fluorophore and the two transition dipoles are not perpendicular, the potential exists for transfer of energy from the first (donor) to the second (acceptor).[70] Thus, if the two fluorophores are close in space, excitation with light at the donor's excitation frequency may result in an emission spectrum characteristic of the acceptor.[71] The most commonly used protein fluorophore pair is cyan fluorescent protein (CFP) and yellow fluorescent protein (YFP). CFP absorbs at around 440 nm and emits maximally at 480 nm, while the emission from YFP can be monitored above 580 nm.

FRET efficiency scales with the inverse sixth power of the distance such that FRET signals occur only if the donor and acceptor are separated by less than 10 nm. In the case of two proteins tagged with donor and acceptor fluorophores, a robust FRET signal therefore strongly implies a genuine binding event. Unfortunately, when using fluorescent proteins, the protein itself occupies much of the useful FRET distance.[72] As a result, FRET in biological systems suffers from a poor signal-to-noise ratio. For example, Chilibeck et al observed a FRET efficiency of 35% (a percentage of the maximal theoretical transfer that would occur at small donor-acceptor distances) for CFP and YFP joined by a ten amino-acid linker and ~18% for CFP/YFP-labelled ADAR enzyme homodimers in HeLa cells.[73] Notably, a further reduction in signal is expected for homo-interactions because only 50% of complexes will contain both donor and acceptor fluorophores.

There are numerous ways of addressing these problems[74] and FRET is still a useful and relatively inexpensive technique for probing complex biological systems. For example, Evans et al used CFP and a commercially-available acceptor ('fluorescein arsenical hairpin') to demonstrate that not only do the two endothelin receptor sub-types (ETA and ETB) form all three possible dimers (A:A, A:B and B:B) but that the heterodimer has a distinct pharmacological response from that of the homodimers.[75]

A slightly different approach, bioluminescence resonance energy transfer (BRET),[76] makes use of an enzyme (typically luciferase) to generate donor light.[77] Because luciferase acts on a substrate in the presence of oxygen, this technique is particularly well-suited to high-throughput or screening assays with live cells in 96-well plates and avoids excitation cross-talk. Hetero- and homo-dimerization of the signalling regulator visual arrestins 2 and 3 was demonstrated using saturating BRET for statistical power and FRET to confirm authentic sub-cellular localization of the interactions.[78] FRET can be further extended by methods that take into account lifetime measurements[79] and anisotropy (below).

FRET with Fluorescence Anisotropy

One of the problems with biological FRET is that overlap between the absorption (and emission) spectra of the donor and acceptor fluorescent proteins causes excitation (and emission) cross-talk. A convenient method to boost signal-to-noise is to measure fluorescence anisotropy.[80] If a GFP donor is excited by a plane polarized light source, the emitted light will retain a high level of polarization because the fluorescence lifetime of GFP is short compared with its rotational correlation time (2.9 ns vs ~20 ns).[80] However, because the donor and acceptor transition dipoles are likely to have different orientations the acceptor emission is likely to be highly depolarized. Therefore any polarized emitted light is mostly due to donor fluorescence and depolarized light can be attributed to acceptor emission.

By the same mechanism, fluorescence anisotropy can be used to detect FRET between two like fluorophores (two molecules of GFP, for example).[81] Steady-state fluorescence anisotropy imaging can detect changes in protein oligomerization in living cells in real time. Using more sophisticated equipment, time-resolved fluorescence anisotropy decay can yield quantitative measurements of molecular rotation time and even information about the separation distance and relative orientation of two dimer subunits.[81]

Proximity Imaging (PRIM)

The PRIM phenomenon relies on the existence of two excitation maxima (395 and 475 nm) for a thermo-tolerant version of GFP.[82] Excitation at the shorter wavelength results in emitted light that is approximately five times as intense as when excited at the longer. This ratio can change when two molecules of GFP are in physical contact,[83] providing an opportunity to directly measure dynamic homo-oligomerization, for example the clustering of glycosylphosphotidylinositide (GPI)-linked membrane proteins in response to proartherogenic stimuli such as oxidized LDL.[84] A major advantage of this scheme is that only one fluorescently-tagged expression construct needs to be made and transfected into cells, making it a simple matter to obtain data rapidly. Care needs to be taken over calibration, although standard fluorescence microscopes and acquisition software can be used.

Fluorescence Recovery after Photobleaching (FRAP)

FRAP[85] is widely used to monitor the diffusion of molecules throughout a cell and trafficking between cellular compartments.[86] Over-excitation of a fluorophore results in a permanent reduction and loss of fluorescent energy emission, i.e., bleaching. When a small, controlled area of a cell expressing a mobile fluorescently-tagged protein is bleached, other tagged molecules will diffuse into the bleached area (and bleached molecules will diffuse out).

Within the cell the rate of observed recovery depends on both the diffusion constant and any association events that occur—particularly to insoluble (effectively immoble) scaffolds. It is essential to determine in a FRAP-based analysis whether or not the observed recovery is diffusion-uncoupled, in which case diffusion occurs on a rapid time-scale compared with binding. In this case there is an initial rapid fluorescence recovery due purely to diffusion followed by a second slower recovery phase in which bleached molecules dissociate from their binding sites and are replaced, hence both diffusion and binding kinetics can be obtained. Alternatively FRAP may be diffusion-coupled whereby the recovery of fluorescence is a function of both binding and diffusion at all time points making component rate parameters more difficult to extract.[86]

FRAP is relatively easy to perform and can be carried out on most confocal microscopes. One difficulty with FRAP lies in the subsequent multi-factorial analysis of the experimental data, which as yet is neither theoretically complete nor straightforward. FRAP may also combined with complementary techniques such as FLIP (fluorescence loss in photobleaching) and IFRAP (inverse FRAP) to identify or eliminate methodological artefacts.[65,87] The use of FRAP is likely to grow as data analysis models become more sophisticated.

In Vivo Complementation

In the yeast two-hybrid system[88] the transcriptional activation and DNA-binding domains of the Gal4 transcriptional activator are separately fused to the two putative protein partners. If the two proteins interact then a functional Gal4 molecule is formed that can selectively bind to the promoter of a reporter gene, and depending on the assay system, activate reporter genes that result in a colour change or a selective advantage on minimal media. A number of comprehensive reviews on this system have been published.[89] To detect homodimerization, the protein of interest is cloned separately behind the two parts of the Gal4 protein and the assay is otherwise carried out in the same manner.[90]

In a recent variation on this method, Stella et al. constructed a system based on the repression of a reporter gene.[91] Expression of the lacZ reporter depends on the activity of a promoter overlapped by two bipartite operator sites. A functional repressor is only formed if the proteins that bind to those sites, P22 and 434, dimerize. By fusing the P22 and 434 DNA-binding domains separately to a putative homodimer it becomes possible to test for homodimerization by looking for a decrease in reporter gene expression. The authors also extended this system to assay for tetramerization.

A number of additional approaches based on fragment complementation have also been described for use in both eukaryotic and bacterial systems. Two otherwise inactive parts of a single enzyme or fluorescent protein are cloned into expression vectors, such that when co-expressed and brought together by interacting fusion proteins, activity is restored.[92] Fragment complementation assays detect the reporter protein itself in any sub-cellular compartment. For example, bimolecular fluorescence complementation (BiFC) uses two nonfluorescent fragments of GFP or one of its variants fused to each subunit of a potential oligomer. Oligomerization results in a reconstituted fluorescent protein, which can be detected using conventional fluorescence microscopy. BiFC complexes are essentially irreversible, so can be used to detect weak or transient interactions, but rarely kinetic parameters. With these caveats in mind, it is possible to use BiFC assays to visualize protein interactions in whole animals, such as Drosophila,[93] Xenopus[94] and nematodes.[95] However, stable BiFC complexes can affect the dynamics of signalling pathways.

β-lactamase fragment complementation yields an active enzyme, which can then process suitable substrates to yield a highly amplified signal. This approach is sensitive and specific and potentially has the capability to detect interactions in vitro with dissociation constants as low as 1 mM.[96] Using the same constructs and a fluorescent substrate, dimerization can be demonstrated in vivo by recording the lactamase-catalysed loss of an intramolecular FRET (see section "Förster Resonance Energy Transfer (FRET)") signal (observed by fluorescence microscopy or in a fluorimetric microtitre plate assay) and then the binding affinity calculated in a colorimetric assay from cell lysates.[97]

CONCLUSION

A number of different approaches are available for the detection and quantitation of protein self-association. Each approach has clear advantages and disadvantages and it is important to consider these issues when assessing the association state of a protein. Generally, it is recommended to make measurements using more than one approach. The combined application of different techniques not only reinforces conclusions drawn from each individual method, but can provide complementary information.

A particular issue is the question of how in vitro measurements correlate with in vivo observations; it is important to establish self-association in both contexts if possible. The introduction of targeted mutations to disrupt dimerization is a recommended control—such an approach can help to establish whether an observed interaction is functionally relevant.

Finally, although it was not discussed above, X-ray crystallography provides detailed structural information about the aggregation state of a protein. However, while these data are seemingly unambiguous, the requirement for efficient packing of individual molecules in the crystal lattice can occasionally give rise to oligomeric structures that are not reproduced in solution. For example, the bacterial superantigen YPM forms a trimer in the crystal state but is monomeric in solution.[98] Thus it is important even in this situation to apply different approaches to have confidence in one's conclusions.

REFERENCES

1. Valdes R Jr, Ackers GK. Study of protein subunit association equilibria by elution gel chromatography. Methods Enzymol 1979; 61:125-142.
2. Siegel LM, Monty KJ. Determination of molecular weights and frictional ratios of proteins in impure systems by use of gel filtration and density gradient centrifugation. Application to crude preparations of sulfite and hydroxylamine reductases. Biochim Biophys Acta 1966; 112:346-362.
3. Cantor CR, Schimmel PR. Biophysical Chemistry. Vol Part II: Techniques for the study of biological structure and function. San Francisco: Freeman, 1980.
4. Andrews P. The gel-filtration behaviour of proteins related to their molecular weights over a wide range. Biochem J 1965; 96:595-606.
5. Stevens FJ, Schiffer M. Computer simulation of protein self-association during small-zone gel filtration. Estimation of equilibrium constants. Biochem J 1981; 195:213-219.
6. Stevens FJ. Analysis of protein-protein interaction by simulation of small-zone size-exclusion chromatography: application to an antibody-antigen association. Biochemistry 1986; 25:981-993.
7. Ceschini S, Lupidi G, Coletta M et al. Multimeric self-assembly equilibria involving the histone-like protein H-NS. A thermodynamic study. J Biol Chem 2000; 275:729-734.
8. Darling PJ, Holt JM, Ackers GK. Coupled energetics of lambda cro repressor self-assembly and site-specific DNA operator binding I: analysis of cro dimerization from nanomolar to micromolar concentrations. Biochemistry 2000; 39:11500-11507.
9. del Alamo M, Neira JL, Mateu MG. Thermodynamic dissection of a low affinity protein-protein interface involved in human immunodeficiency virus assembly. J Biol Chem 2003; 278:27923-27929.
10. van Holde KE, Johnson WC, Ho PS. Principles of Physical Biochemistry. 2 ed. Upper Saddle River: Prentice-Hall, 2006.
11. Laue TM, Bhairavi DS, Ridgeway TM et al. Computer-aided interpretation of analytical sedimentation data. In: Harding SE, Rowe AJ, Horton JC, eds. Analytical Ultracentrifugation in Biochemistry and Polymer Science. Cambridge: Royal Society of Chemistry, 1992:90-125.
12. Rowe AJ. The concentration dependence of sedimentation. In: Harding SE, Rowe AJ, Horton JC, eds. Analytical Ultracentrifugation in Biochemistry and Polymer Science. Cambridge: Royal Society of Chemistry, 1992.
13. Demeler B, Saber H, Hansen JC. Identification and interpretation of complexity in sedimentation velocity boundaries. Biophys J 1997; 72:397-407.
14. Stafford WF. Sedimentation boundary analysis of interactiong systems: use of the apparent sedimentation coefficient distribution function. In: Shuster TM, Laue TM, eds. Modern Analytical Ultracentrifugation. Boston: Birkhauser, 1994:119-137.
15. Cole JL. Characterization of human cytomegalovirus protease dimerization by analytical centrifugation. Biochemistry 1996; 35:15601-15610.
16. Ali SA, Iwabuchi N, Matsui T et al. Reversible and fast association equilibria of a molecular chaperone, gp57A, of bacteriophage T4. Biophys J 2003; 85:2606-2618.
17. Zhao H, Beckett D. Kinetic partitioning between alternative protein-protein interactions controls a transcriptional switch. J Mol Biol 2008; 380:223-236.
18. Nan R, Gor J, Perkins SJ. Implications of the progressive self-association of wild-type human factor H for complement regulation and disease. J Mol Biol 2008; 375:891-900.
19. Waxman E, Laws WR, Laue TM et al. Refining hydrodynamic shapes of proteins: the combination of data from analytical ultracentrifugation and time-resolved fluorescence anisotropy decay. In: Shuster TM, Laue TM, eds. Modern Analytical Ultracentrifugation. Boston: Birkhauser, 1994:189-205.
20. Burian J, Ausio J, Phipps B et al. Hexamerization of RepA from the Escherichia coli plasmid pKL1. Biochemistry 2003; 42:10282-10287.
21. Fleming KG, Hohl TM, Yu RC et al. A revised model for the oligomeric state of the N-ethylmaleimide-sensitive fusion protein, NSF. J Biol Chem 1998; 273:15675-15681.
22. Lebowitz J, Lewis MS, Schuck P. Modern analytical ultracentrifugation in protein science: a tutorial review. Protein Sci 2002; 11:2067-2079.
23. Gell D, Kong Y, Eaton SA et al. Biophysical characterization of the alpha-globin binding protein alpha-hemoglobin stabilizing protein. J Biol Chem 2002; 277:40602-40609.
24. McRorie DK, Voelker PJ. Self-associating systems in the analytical ultracentrifuge. Beckman Instruments, Fullerton, CA: Beckman Instruments, Fullerton, CA, 1993.
25. Folta-Stogniew E, Williams KR. Determination of molecular masses of proteins in solution: implementation of an HPLC size exclusion chromatography and laser light scattering service in a core laboratory. J Biomol Tech 1999; 10:51-63.
26. Wyatt PJ. Light scattering and the absolute characterization of macromolecules. Anal Chim Acta 1993; 272:1-40.
27. Philo JS, Aoki KH, Arakawa T et al. Dimerization of the extracellular domain of the erythropoietin (EPO) receptor by EPO: one high-affinity and one low-affinity interaction. Biochemistry 1996; 35:1681-1691.

28. White JF, Grodnitzky J, Louis JM et al. Dimerization of the class A G protein-coupled neurotensin receptor NTS1 alters G protein interaction. Proc Natl Acad Sci USA 2007; 104:12199-12204.
29. Philo JS. Is any measurement method optimal for all aggregate sizes and types? AAPS J 2006; 8:E564-571.
30. Papish AL, Tari LW, Vogel HJ. Dynamic light scattering study of calmodulin-target peptide complexes. Biophys J 2002; 83:1455-1464.
31. Khanova HA, Markossian KA, Kurganov BI et al. Mechanism of chaperone-like activity. Suppression of thermal aggregation of betaL-crystallin by alpha-crystallin. Biochemistry 2005; 44:15480-15487.
32. Habel JE, Ohren JF, Borgstahl GEO. Dynamic light-scattering analysis of full-length human RPA14/32 dimer: purification, crystallization and self-association. Acta Crystallogr D 2001; 57:254-259.
33. Lipfert J, Doniach S. Small-angle X-ray scattering from RNA, proteins and protein complexes. Annu Rev Biophys Biomol Struct 2007; 36:307-327.
34. Koch MH, Vachette P, Svergun DI. Small-angle scattering: a view on the properties, structures and structural changes of biological macromolecules in solution. Q Rev Biophys 2003; 36:147-227.
35. Jánosi A. The exponential distribution in small angle X-ray scattering. Theory and practice. Monatshefte für Chemie/Chemical Monthly 1993; 124:815-826.
36. Rescic J, Vlachy V, Jamnik A et al. Osmotic Pressure, small-angle X-ray and dynamic light scattering studies of human serum albumin in aqueous solutions. J Colloid Interf Sci 2001; 239:49-57.
37. Bada M, Walther D, Arcangioli B et al. Solution structural studies and low-resolution model of the Schizosaccharomyces pombe sap1 protein. J Mol Biol 2000; 300:563-574.
38. Whitten AE, Jacques DA, Hammouda B et al. The structure of the KinA-Sda complex suggests an allosteric mechanism of histidine kinase inhibition. J Mol Biol 2007; 368:407-420.
39. Schroder E, Willis AC, Ponting CP. Porcine natural-killer-enhancing factor-B: oligomerisation and identification as a calpain substrate in vitro. Biochim Biophys Acta 1998; 1383:279-291.
40. Tang KH, Guo H, Yi W et al. Investigation of the conformational states of Wzz and the Wzz.O-antigen complex under near-physiological conditions. Biochemistry 2007; 46:11744-11752.
41. Velazquez-Campoy A, Leavitt SA, Freire E. Characterization of protein-protein interactions by isothermal titration calorimetry. Methods Mol Biol 2004; 261:35-54.
42. Burrows SD, Doyle ML, Murphy KP et al. Determination of the monomer-dimer equilibrium of interleukin-8 reveals it is a monomer at physiological concentrations. Biochemistry 1994; 33:12741-12745.
43. Schulte A, Czudnochowski N, Barboric M et al. Identification of a cyclin T-binding domain in Hexim1 and biochemical analysis of its binding competition with HIV-1 Tat. J Biol Chem 2005; 280:24968-24977.
44. McPhail D, Cooper A. Thermodynamics and kinetics of dissociation of ligand-induced dimers of vancomycin antibiotics. J Chem Soc, Faraday Trans 1997; 93:2283-2289.
45. Dingley AJ, Mackay JP, Chapman BE et al. Measuring protein self-association using pulsed-field-gradient NMR spectroscopy: application to myosin light chain 2. J Biomol NMR 1995; 6:321-328.
46. Baden HA, Sarma SP, Kapust RB et al. The amino-terminal domain of human STAT4. Overproduction, purification and biophysical characterization. J Biol Chem 1998; 273:17109-17114.
47. Nilges M. A calculation strategy for the structure determination of symmetric dimers by 1H NMR. Proteins 1993; 17:297-309.
48. Lee W, Harvey TS, Yin Y et al. Solution structure of the tetrameric minimum transforming domain of p53. Nat Struct Biol 1994; 1:877-890.
49. O'Donoghue SI, Chang X, Abseher R et al. Unraveling the symmetry ambiguity in a hexamer: calculation of the R6 human insulin structure. J Biomol NMR 2000; 16:93-108.
50. Folkers PJM, Folmer RHA, Konings RNH et al. Overcoming the ambiguity problem encountered in the analysis of nuclear overhauser magnetic resonance spectra of symmetric dimer proteins. J Am Chem Soc 1993; 115:3798-3799.
51. Otting G, Wüthrich K. Extended heteronuclear editing of 2D 1H NMR spectra of isotope-labeled proteins, using the X([omega]1, [omega]2) double half filter. J Magn Reson 1989; 85:586-594.
52. Drohat AC, Amburgey JC, Abildgaard F et al. Solution structure of rat apo-S100B(beta beta) as determined by NMR spectroscopy. Biochemistry 1996; 35:11577-11588.
53. Jasanoff A, Fersht AR. Quantitative determination of helical propensities from trifluoroethanol titration curves. Biochemistry 1994; 33:2129-2135.
54. Light-Wahl KJ, Schwartz BL, Smith RD. Observation of the noncovalent quaternary associations of proteins by electrospray ionization mass spectrometry. J Am Chem Soc 1994; 116:5271-5278.
55. Keetch CA, Bromley EH, McCammon MG et al. L55P transthyretin accelerates subunit exchange and leads to rapid formation of hybrid tetramers. J Biol Chem 2005; 280:41667-41674.
56. Rostom AA, Robinson CV. Detection of the intact GroEL chaperonin assembly by mass spectrometry. J Am Chem Soc 1999; 121:4718-4719.
57. Hernandez H, Robinson CV. Determining the stoichiometry and interactions of macromolecular assemblies from mass spectrometry. Nat Protocols 2007; 2:715-726.
58. Jameson DM, Seifried SE. Quantification of protein-protein interactions using fluorescence polarization. Methods 1999; 19:222-233.

59. Maleki SJ, Royer CA, Hurlburt BK. MyoD-E12 heterodimers and MyoD-MyoD homodimers are equally stable. Biochemistry 1997; 36:6762-6767.
60. Ryan DP, Duncan JL, Lee C et al. Assembly of the oncogenic DNA-binding complex LMO2-Ldb1-TAL1-E12. Proteins 2008; 70:1461-1474.
61. de Lumley M, Hart DJ, Cooper MA et al. A biophysical characterisation of factors controlling dimerisation and selectivity in the NF-kappaB and NFAT families. J Mol Biol 2004; 339:1059-1075.
62. Bakkenist CJ, Kastan MB. DNA damage activates ATM through intermolecular autophosphorylation and dimer dissociation. Nature 2003; 421:499-506.
63. Hillar A, Culham DE, Vernikovska YI et al. Formation of an antiparallel, intermolecular coiled coil is associated with in vivo dimerization of osmosensor and osmoprotectant transporter ProP in Escherichia coli. Biochemistry 2005; 44:10170-10180.
64. Westphal V, Rizzoli SO, Lauterbach MA et al. Video-rate far-field optical nanoscopy dissects synaptic vesicle movement. Science 2008; 320:246-249.
65. Kimura H, Hieda M, Cook PR. Measuring histone and polymerase dynamics in living cells. Methods Enzymol 2003; 375:381-393.
66. Bacia K, Schwille P. A dynamic view of cellular processes by in vivo fluorescence auto- and cross-correlation spectroscopy. Methods 2003; 29:74-85.
67. Philip F, Sengupta P, Scarlata S. Signaling through a G protein-coupled receptor and its corresponding G protein follows a stoichiometrically limited model. J Biol Chem 2007; 282:19203.
68. Kask P, Palo K, Ullmann D et al. Fluorescence-intensity distribution analysis and its application in biomolecular detection technology. Proc Natl Acad Sci USA 1999; 96:13756-13761.
69. Saffarian S, Li Y, Elson EL et al. Oligomerization of the EGF receptor investigated by live cell fluorescence intensity distribution analysis. Biophys J 2007; 93:1021-1031.
70. Förster T. Zwischenmolekulare energiewanderung und fluoreszenz. Ann Phys 1948; 437:55-75.
71. Sapsford KE, Berti L, Medintz IL. Materials for fluorescence resonance energy transfer analysis: beyond traditional donor-acceptor combinations. Angew Chem Int Ed Engl 2006; 45:4562-4589.
72. Rizzo MA, Springer G, Segawa K et al. Optimization of pairings and detection conditions for measurement of FRET between cyan and yellow fluorescent proteins. Microsc Microanal 2006; 12:238-254.
73. Chilibeck KA, Wu T, Liang C et al. FRET analysis of in vivo dimerization by RNA-editing enzymes. J Biol Chem 2006; 281:16530-16535.
74. Piston DW, Kremers GJ. Fluorescent protein FRET: the good, the bad and the ugly. Trends Biochem Sci 2007; 32:407-414.
75. Evans NJ, Walker JW. Endothelin receptor dimers evaluated by FRET, ligand binding and calcium mobilization. Biophys J 2008; 95:483-492.
76. Bacart J, Corbel C, Jockers R et al. The BRET technology and its application to screening assays. Biotechnol J 2008; 3:311-324.
77. Gandiá J, Lluis C, Ferre S et al. Light resonance energy transfer-based methods in the study of G protein-coupled receptor oligomerization. Bioessays 2008; 30:82-89.
78. Storez H, Scott MG, Issafras H et al. Homo- and hetero-oligomerization of beta-arrestins in living cells. J Biol Chem 2005; 280:40210-40215.
79. Clayton AH, Walker F, Orchard SG et al. Ligand-induced dimer-tetramer transition during the activation of the cell surface epidermal growth factor receptor-A multidimensional microscopy analysis. J Biol Chem 2005; 280:30392-30399.
80. Piston DW, Rizzo MA. FRET by fluorescence polarization microscopy. Methods Cell Biol 2008; 85:415-430.
81. Gautier I, Tramier M, Durieux C et al. Homo-FRET microscopy in living cells to measure monomer-dimer transition of GFP-tagged proteins. Biophys J 2001; 80:3000-3008.
82. Siemering KR, Golbik R, Sever R et al. Mutations that suppress the thermosensitivity of green fluorescent protein. Curr Biol 1996; 6:1653-1663.
83. De Angelis DA, Miesenbock G, Zemelman BV et al. PRIM: proximity imaging of green fluorescent protein-tagged polypeptides. Proc Natl Acad Sci USA 1998; 95:12312-12316.
84. Patschan S, Li H, Brodsky S et al. Probing lipid rafts with proximity imaging: actions of proatherogenic stimuli. Am J Physiol Heart Circ Physiol 2006; 290:H2210-2219.
85. Sprague BL, McNally JG. FRAP analysis of binding: proper and fitting. Trends Cell Biol 2005; 15:84-91.
86. Meyer T, Begitt A, Vinkemeier U. Green fluorescent protein-tagging reduces the nucleocytoplasmic shuttling specifically of unphosphorylated STAT1. Febs J 2007; 274:815-826.
87. Stavreva DA, McNally JG. Fluorescence recovery after photobleaching (FRAP) methods for visualizing protein dynamics in living mammalian cell nuclei. Methods Enzymol 2004; 375:443-455.
88. Fields S, Song O. A novel genetic system to detect protein protein interactions. Nature 1989; 340:245-246.
89. MacDonald PN, ed. Two-Hybrid Systems: Methods and Protocols. New York: Springer-Verlag; 2001; No. 177.
90. Tsai WC, Pan ZJ, Hsiao YY et al. Interactions of B-class complex proteins involved in tepal development in Phalaenopsis orchid. Plant Cell Physiol Vol 49; 2008:814-824.
91. Stella S, Spurio R, Falconi M et al. Nature and mechanism of the in vivo oligomerization of nucleoid protein H-NS. EMBO J 2005; 24:2896-2905.

92. Villalobos V, Naik S, Piwnica-Worms D. Current state of imaging protein-protein interactions in vivo with genetically encoded reporters. Annu Rev Biomed Eng 2007; 9:321-349.
93. Benton R, Sachse S, Michnick SW et al. Atypical membrane topology and heteromeric function of Drosophila odorant receptors in vivo. PLoS Biol 2006; 4:e20.
94. Saka Y, Hagemann AI, Piepenburg O et al. Nuclear accumulation of Smad complexes occurs only after the midblastula transition in Xenopus. Development 2007; 134:4209-4218.
95. Shyu YJ, Hiatt SM, Duren HM et al. Visualization of protein interactions in living Caenorhabditis elegans using bimolecular fluorescence complementation analysis. Nat Protoc 2008; 3:588-596.
96. Magliery TJ, Wilson CG, Pan W et al. Detecting protein-protein interactions with a green fluorescent protein fragment reassembly trap: scope and mechanism. J Am Chem Soc 2005; 127:146-157.
97. Galarneau A, Primeau M, Trudeau LE et al. Beta-lactamase protein fragment complementation assays as in vivo and in vitro sensors of protein protein interactions. Nat Biotechnol 2002; 20:619-622.
98. Donadini R, Liew CW, Kwan AH et al. Crystal and solution structures of a superantigen from Yersinia pseudotuberculosis reveal a jelly-roll fold. Structure 2004; 12:145-156.
99. Johnson ML, Correia JJ, Yphantis DA et al. Analysis of data from the analytical ultracentrifuge by nonlinear least-squares techniques. Biophys J 1981; 36:575-588.
100. Junius FK, Mackay JP, Bubb WA et al. Nuclear magnetic resonance characterization of the Jun leucine zipper domain: unusual properties of coiled-coil interfacial polar residues. Biochemistry 1995; 34:6164-6174.

CHAPTER 3

COMPUTATIONAL AND STRUCTURAL CHARACTERISATION OF PROTEIN ASSOCIATIONS

Susan Jones
Department of Biochemistry, School of Life Sciences, University of Sussex, Brighton, UK
Email: s.jones@sussex.ac.uk

Abstract: Protein-protein associations represent the building blocks of biological systems. The classification of different types of protein association is fundamental to an understanding of the interactions they exhibit. A protein association can be classified as homo- (identical components) or hetero- (non-identical components) and in addition permanent (components only exist and function in an associated state) or transient (components exist independently but interact for a limited time to carry out a specific function). A large number of studies have analysed the physical and chemical characteristics of protein-protein interactions using three-dimensional structures derived from X-ray crystallography. This chapter summarises the major conclusions of these studies, focusing on amino acid preferences and secondary structure packing at interfaces: hydration, hydrophobic and electrostatic effects, conformational changes and evolutionary conservation. The studies highlight differences between the interaction sites and the rest of the protein surface and between different classes of protein association. Common themes in the interfaces of protein associations are also revealed including shape complementarity, the presence of water molecules, a high percentage of arginine residues, intermolecular hydrogen bonds and an energy of association comprising hydrophobic and electrostatic effects. These studies also emphasise how the relative importance of such characteristics is dependant upon the class of protein association, with permanent associations generally displaying different characteristics to transient associations.

INTRODUCTION

Protein associations represent the building blocks of biological systems. Knowledge of the nature of protein associations has increased in parallel with the number of three

Protein Dimerization and Oligomerization in Biology, edited by Jacqueline M. Matthews.

dimensional protein structures available. This large volume of structural data has fuelled computational research focussed on the analysis of the physical and chemical characteristics of the interactions within these associations. With a greater knowledge of protein associations it will be possible to predict which proteins interact and build interaction networks for biochemical pathways and whole organisms. Hence this area of research underpins the emerging field of systems biology.

The role of protein association is to achieve a structure that is energetically and functionally favourable. This requires interaction between two surfaces that are complementary in terms of shape and chemistry. How this complementarity is achieved varies for different types of protein association and this chapter begins with an introduction to the classification of protein associations, which is fundamental to an understanding of the characteristics they exhibit.[a] Then a discussion follows on the structural and chemical properties of interaction sites that have been derived from studies on protein structures extracted from the Protein Data Bank (PDB).[1] This discussion focuses on amino acid preferences, secondary structure packing, hydration, hydrophobic and electrostatic effects, conformational changes and evolutionary conservation. The chapter concludes with a brief summary of how knowledge of protein-protein interactions will develop in the future.

CLASSIFICATION OF PROTEIN-PROTEIN COMPLEXES

As the number of dimeric protein associations has increased so the need to classify them into functionally and structurally related classes has become evident. A dimeric association is either between components that are identical, termed a homodimer, or between non-identical components, termed a heterodimer. At a further level of classification the permanence of a complex and its functional significance can also be considered. Classification at this level becomes confusing, partly due the duplicity of terminology used in the literature. Jones and Thornton distinguish between obligatory interactions and transient interactions, the former being identified as those that are in permanent contact and the latter as those between separate proteins interacting for a limited time to carry out a specific function.[2] In later work Nooren and Thornton define complexes as either obligate or non-obligate and additionally either permanent or transient (Fig. 1A).[3] An obligate complex is defined as one in which the individual subunits are not found as stable structures in vivo and non-obligate as those with subunits that exist independently. The terms permanent and transient interactions attempt to classify the lifetime of a complex, with permanent complexes being classified as very stable associations whilst transient complexes have subunits that associated and dissociated in vivo. In this classification obligate complexes can only be permanent but non-obligate complexes can be permanent or transient.[3]

Defining classes can be misleading as many protein associations can be classified into more than one class. In vivo a continuum exists between obligate and non-obligate interactions and the longevity of an interaction is dependent upon the conditions within the cell.[3] In the literature the paired terms obligate: non-obligate and permanent: transient have been used interchangeably and both continue to be used to classify protein-protein complexes.[4,5] Ofran and Rost[6] have also introduced the term obligomer to define interactions between subunits that are obligatory as defined by Jones and Thornton.[2] Ofran and Rost[6] actually define four classes of interface relevant to this discussion (1) homo-obligomer: interactions between permanently interacting identical subunits (2) homo-complex: interactions between transiently interacting chains (3) hetero-obligomer: interactions between permanently interacting different protein subunits and (4) hetero-complex: interactions between transiently interacting different protein subunits (Fig. 1B). Examples of each of the four classes of protein-protein association are shown in Figure 2 and Table 1. The introduction of this new terminology helps to clarify classifications and the terms complex (transient homo and hetero interactions) and obligomer (permanent homo and hetero interactions) will be adopted in this chapter. In addition the term protein

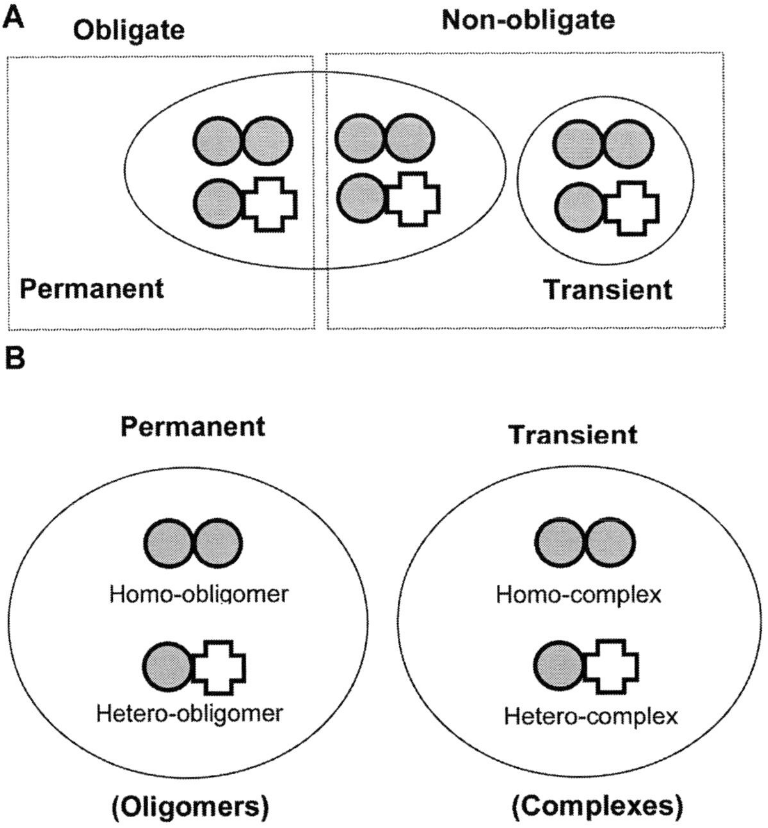

Figure 1. Classification of protein associations. Homo interactions (between identical subunits) are depicted by a pair of identical symbols. Hetero interactions (between non-identical subunits) are depicted by a pair of non-identical symbols. A) Classification derived by Nooren and Thornton in which a complex can be defined as obligate or non-obligate and in addition permanent or transient.[3] B) Classification derived by Ofran and Rost in which four classes of interaction, relevant to this chapter, are defined and the term obligomer is introduced.[6]

Figure 2, viewed on following page. Examples of the 4 classes of protein-protein associations deposited in the PDB.[1] A) Homo-obligomer: Cardiotoxin from *Naja mossambica* (PDB code 1cdt).[52] B) Hetero-obligomer: Chorionic gonadotrophin from *Homo sapiens* (PDB code 1hrp).[53] C) Homo-complex: Lysin dimer from *Haliotis rufescens* (PDB code 2lyn)[54] and D) Hetero-complex: EF-Tu.EF-Ts complex from *E. coli* (PDB code 1efu).[55] In each figure one part of the protein association is shown in white and the second in grey. The figures on the left depict the associations as CPK models and on the right as secondary structure cartoons. See Table 1 for more information relating to these complexes.

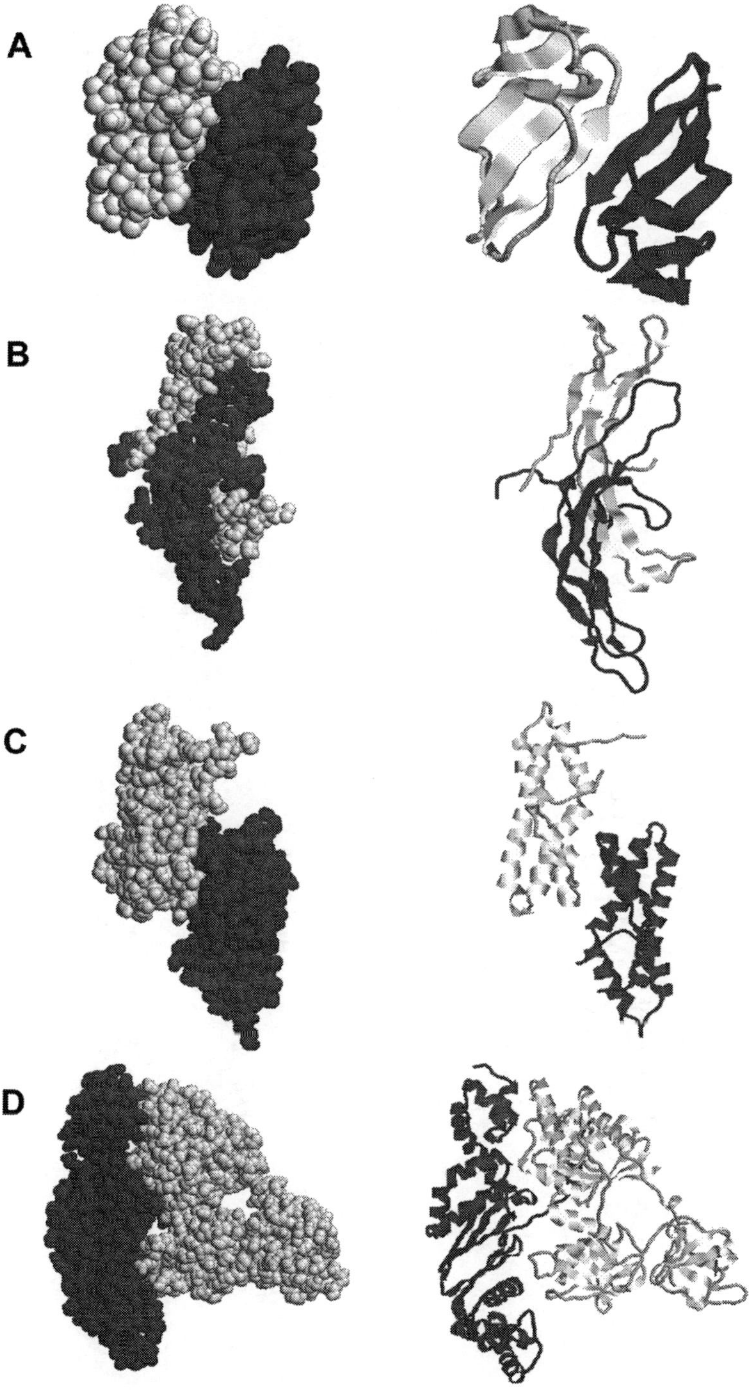

Figure 2. Please see figure legend on previous page.

Table 1. A Summary of the Physical and Chemical Characteristics of Interfaces in Protein Associations. An Example of each of the 4 Classes of Association are Named (a-c); but the Interface Characteristics Shown are Generic for Either a Specific Class of Association (d-f) or for all Classes of Association (g-i).

	Obligomers		Complexes	
	Homo	Hetero	Homo	Hetero
(a) Example	Cardiotoxin Naja mossambica	Chorionic gonadotrophin Homo sapiens	Lysin dimer Haliotis rufescens	EF-Tu.EF-Ts complex E.coli
(b) PDB[1] Code	1cdt	1hrp	2lyn	1efu
(c) Figure	2a	2b	2c	2d
(d) K_d M	---	---	1×10^{-6}	3×10^{-8}
(e) Favoured Amino Acids	>> Hydrophobic	>> Hydrophobic	> Hydrophobic	>> Polar
(f) (c) Hydration (waters per 100 $Å^2$ interface)	1.13[27]	---	---	1.0[13,27]
(g) (d) Secondary Structure	Like-with-like packing e.g helix-helix & strand-strand. Turns and loops over represented.[5,26]			
(h) Amino acids in interface core regions	Tryptophan and tyrosine[5,6,17,19]			
(i) Conformational changes	(i) Disorder-to-order transitions, (ii) large movements of loop structures (iii) domain movements[13]			

association will also be used to define all types of protein-protein interaction, including obligomers and complexes.

This discussion on classification applies to complexes that are of biological relevance, i.e., the protein association has a biological function within the organism. However, when a protein structure is solved using X-ray crystallography the asymmetric unit of the crystal can contain additional contacts between protein components that are not biologically relevant.[7] For example, a protein may function as a biological monomer in a cell but be crystallized as a dimer. These interactions are artefacts of crystal packing and they exhibit different characteristics to those of biological relevance.[8] The differences observed in terms of accessible surface area buried in the interaction and the chemical complementarity of the interface have been used to predict the biological quaternary state of a protein.[7,9] However, this chapter will now focus on protein associations that are of biological relevance.

CHARACTERISATION OF PROTEIN-PROTEIN ASSOCIATIONS

The analysis of interfaces in protein associations began as soon as the first structures were deposited in the PDB; with Chothia and Janin publishing an analysis of just 3 complexes, the insulin dimer, trypsin-PTI complex and αβ oxyhaemoglobin dimer in 1975.[10] This study assessed the size of the accessible surface area buried upon association, the amino acid preferences observed in the interface and the number of intermolecular hydrogen bonds.[10] This led to a series of structural analyses on increasingly large data sets extracted from the PDB that are now discussed.[2,4-6,8,11-19]

Amino Acid Preferences

It is well known that there are differences in the distribution of amino acids between the interior and exterior of a protein.[20] In addition studies have shown that the distribution of amino acids in a protein-protein interface is different to that of the rest of the protein surface.[13,14,16,21] However, as noted by Ofran and Rost,[6] some studies appear to be contradictory in terms of the specific amino acid preferences shown at interface sites. Some conclude that the distribution of amino acids in the interface is intermediate of that observed in the protein interior and the exterior,[13] whilst others have shown that the distribution is more similar to the protein interior.[11] In addition some analyses show that polar and charged residues make major contributions to interactions,[22,23] whilst others conclude that polar residues do not play key roles.[24] The contradictory nature of these results can be explained if the class of protein association analysed is considered. Some analyses are based on data sets of one class of association e.g., homo-obligomers,[14] hetero-complexes[13] or homo and hetero complexes.[5,25] Whilst other analyses have been based on larger datasets comprising varying numbers of homo and hetero obligomers and complexes.[2,4,6,8,16-18]

Glaser et al[18] studied a mixed data set of 621 protein associations and found that larger interfaces (those >5000 Å^2 accessible surface area buried for both parts of the association) were more hydrophobic than smaller interfaces which featured more polar residues. In this paper the data set was not classified into permanent or transient associations and hence the conclusion was that the amino acid preferences are similar for different classes of protein association. In contrast two other studies concluded that there are significant differences in the amino acid composition of the interface between different classes of protein association.[2,13] Analysis of 58 obligomers and complexes revealed that in general hydrophobic residues have a greater preference for the interfaces of homodimers than those of hetero-complexes, with the hetero-complex interfaces balanced by an increased number of polar residues.[2] Similar conclusions were drawn from a data set of 75 hetero-complexes and 23 obligomers that showed, the composition of the obligomer interfaces being much closer to that of the protein interior, exhibiting more hydrophobic residues and fewer charged residues.[13] The analysis of Ofran and Rost,[6] which divided interfaces into 4 types relevant to this discussion (Fig. 1B), found that large hydrophobic residues including histidine, methionine and tyrosine, were favoured in all types of interaction and that hydrophobic residues were more common in homo than hetero interactions.

The amino acid preferences revealed in different classes of protein associations can be explained in terms of energetics. Components of transient associations cannot have large hydrophobic patches on their surfaces because they also exist independently. In the independent state exposed hydrophobic patches on the protein surface would be energetically unfavourable. Hence, such associations feature more polar interactions. In contrast, obligomer interfaces are permanently shielded from the solvent by the second protein component of the association and hence hydrophobic patches on the surface of these protein components are not energetically unfavourable.

A further complicating factor is that residues appear to be favoured or disfavoured in different parts of the interface. A number of groups have divided interfaces into at least two areas, referred to here as the core and the rim.[4,14,17,19] Borgan and Thorn identified core regions or 'hot spots' in binding energies that had high frequencies of tryptophan, tyrosine and arginine residues.[17] Other analyses have also shown that tryptophan and tyrosine are favoured residues in the core of the interface.[5,6,19] It has been suggested that tyrosine is favoured as it has few rotatable bonds meaning it can contribute to the energy of association through the hydrophobic effect without a negative energy component.[19]

SECONDARY STRUCTURE PACKING

The secondary structure composition of interfaces has also been analysed.[5,12,14] Early work on a small data set of 19 dimers and tetramers revealed interface contacts between different types of secondary structure including helix-helix packing, β-sheet packing, extended β-sheets and loop interactions.[12] In an analysis of 170 nonredundant complexes it was found that helices and sheets occurred less frequently than expected and that turns and loop structures were over presented.[5] The presence of turns and loops in interface is likely to create a flexible environment that will be advantageous for tight packing. A preference for like-with-like secondary structure packing, such as helix-helix and sheet-sheet packing has also been observed by a number of groups.[5,26] Again it is likely that such secondary structure pairings afford more opportunities for tight packing than non-identical secondary structure packing, thus contributing to the complementarity of the interface.

HYDRATION

Hydration of protein-protein interfaces is an important factor that contributes to tight packing and the energy of association. However, the analysis of the number of water molecules and the patterns of interaction they make is limited by the lack of consistency in the way in which the position of solvent molecules are reported in crystallographic structures.[27] In general there is more reliable reporting of solvent molecules in protein crystals with higher resolutions and hence this tends to limit the size of data sets that can be studied. More recently an algorithm that can predict the number and location of water molecules within protein associations has been developed and can be applied to structures for which experimental data is missing or incomplete.[28]

An analysis of a dataset of homo-obligomers and complexes solved to a minimum of 2.6 Å resolution revealed that interfaces have on average 1.13 and 1.0 water molecules per 100 Å^2 respectively.[28] A similar figure of 1.0 interface water molecule per 100 Å^2 of interface accessible surface area was also observed for a data set of hetero-complexes.[13] Many of these interface water molecules form hydrogen bonds with both components of the association and it is proposed that protein-protein associations contain at least as many water mediated contacts as direct contacts.[29] The distribution of water molecules within the interface varies widely, with some interfaces observed to have solvent molecules distributed around the edge of the interface leaving the centre "dry", whilst others are defined as "wet", with water molecules distributed throughout.[27] In many of the "wet" interfaces water molecules occupy cavities which has the effect of increasing the close packing of the association and hence the complementarity.[13,30]

Hydrophobic Forces

The hydrophobic effect is the property that causes nonpolar molecules to self-associate in an aqueous environment, which results in a gain in free energy.[31] This effect is considered by many to be the primary driving force in the stabilisation of protein associations,[10,32] but its contribution to this process is not believed to be as large as its contribution to protein folding.[33,34] At present methods for the calculation of the hydrophobic effect give differing values and the relative contribution of hydrophobic forces to the energy of association is still disputed. Based on the analysis of amino acid preferences in protein associations it is likely that the relative contribution will be dependant upon the class of association.

As previously highlighted, protein interfaces can be subdivided into core and rim regions that display different amino acid preferences.[4,14,17,19] Alanine scanning mutagenesis studies have shown that clusters of residues or "hot spots" in the central zone of the interface dominate the energetics of protein association.[13,17] This observation led to the O-ring hypothesis which infers that a condition for high affinity binding between two proteins is the exclusion of solvent from the interface.[17] This solvent exclusion is achieved by a ring of residues surrounding the central "hot spots" that are largely hydrophobic.[17] The hydrophobic contribution in each of the interface regions has been measured experimentally for the interface of antigen-antibody complexes using microcalorimetry.[35,36] From these experiments the hydrophobicity of the O-ring region was estimated to be 21 cal mol^{-1} for the burial of 1 Å^2 of apolar surface[35] and 46 cal mol^{-1} for the burial of 1 Å^2 of apolar surface for the central region.[36] The higher value estimated for the central region agrees with some of the estimates of the hydrophobic effect in the cores of protein structure. e.g.,[37] Hence it is likely that there is no single measure of the hydrophobic effect in a protein-protein interface, but rather a continuum of values dependent upon the position within the interface; and the overall effect is dependent upon the class of association.

Electrostatic Forces

The analysis of amino acid preferences in some protein-protein interfaces has revealed large numbers of charged and polar residues that are either fully or partially buried in the interface.[22,23] Many of these residues are involved in hydrogen bonding and salt bridges.[14] The energetic contribution of electrostatic interactions in protein associations can be considered as the balance of two effects: the negative effect of the loss of favourable interactions between polar and charged residues and solvent; and, the positive effect of the formation of new hydrogen bonds and ion pairs within the interface.[38] It has been considered that the positive effect of the latter is not usually large enough to compensate for the loss of favourable solvent contacts,[39-41] and hence electrostatic interactions have a negative effect on protein associations.[38]

However, theoretical calculations of the contribution of electrostatic interactions in four complexes showed that the contribution of these interactions varies widely between different complexes.[38] Two of the complexes studied, human growth hormone complexed with its receptor and the subtype N9 influenza virus neuramidase complexed with the NC41 antibody, involved electrostatic interactions that strongly opposed binding. In a third complex, barnase-barstar, the net effect of the electrostatic interactions was close to zero and in the fourth complex (the Ras binding domain of kinase cRaf and Rap1A) these interactions actually favoured binding. This work also found that the total electrostatic contribution to binding is inversely correlated to the buried total and nonpolar surface: i.e., the hydrophobic effect.[38]

Hence the relative contributions of electrostatic and hydrophobic effects to the energy of association appear to dependant upon the type of association. However, one parameter that is often excluded from energy calculations is the energy loss or gain from conformational changes that can occur upon protein association. Some models can take into account small side chain rearrangements[42] but the effect of large conformational changes on the energetics of protein has yet to be calculated and hence is unclear.

CONFORMATIONAL CHANGES

As mentioned above many proteins undergo conformational changes when they associate. Such changes can range from small changes in side chain conformations of interface amino acids to large changes in the orientation of whole domains. Conformational changes occur for a number of reasons including: the facilitation of close packing; orientation of residues for a specific function; and, the formation of intermolecular hydrogen bonds or salt bridges.[43]

The analysis of conformational changes relies on a protein's structure being available in both the associated and un-associated state; and whilst databases of such structures are growing,[44] they are still too small to make many conclusions. In addition, methods for the comparison of such pairs of structures are also limited, some relying on structural superposition of structures and the calculation of root mean square deviations,[45] whereas others use distance matrices.[46] The difficulty with comparing proteins from two different crystal structures is to find which changes are as a result of differences in crystal packing and experimental conditions, and which are as a result of protein association.

Three major types of structural changes have been identified in a dataset of 75 hetero associations (1) disorder-to-order transitions (2) large movements of loop structures and

(3) changes in the relative orientation of the domains.[13] The number and type of changes that occur vary widely and it can be that just one, or both, component(s) of the association undergoe(s) a conformational change. In addition several types of change may be observed within one protein association, for example loop movements may take place at the same time as domain movements. Analyses have shown that conformational changes required to orient parts of the protein for a specific function are often large compared to those that aid the creation of a compact interface or the creation of intermolecular bonds.[45]

EQUILIBRIUM CONSTANTS

As discussed above, protein-protein association is the result of a number of different forces (including hydrophobic and electrostatic forces) and additional elements such as hydration and conformational changes. To evaluate the strengths and gauge the lifetimes of protein associations researchers have used dissociation constants. A dissociation constant (K_d) is a specific class of equilibrium constant that gives a measure of the propensity for a two entities (in this case proteins) to separate into two components. The smaller the K_d the more tightly bound the two proteins in the association. The two complexes shown in Figure 2, the homo complex lysine (Fig. 2C) and the hetero complex EF-Tu.EF-Ts (Fig. 2D) have reported values of K_d of 1×10^{-6} M and 3×10^{-8} M respectively.[25] Whilst the evaluation of K_d data for protein associations is useful for summarizing the kinetics, it has proved experimentally difficult to obtain such values for many associations. In addition, the K_d for a single protein association can be dependant upon experimental conditions, such as temperature. However, in recent years the volume of K_d data available for protein associations has grown and those reported in the literature are stored in a number of databases.[47,48] The availability of K_d data for complexes has revealed a linear relationship between K_d and the size of the contact site, with interfaces of high affinity complexes being larger than low affinity complexes.[25] However, the K_d data for obligomers is still very limited; and until such data is more widely available for all types of protein associations, a valid comparison of the kinetics of the different classes of association is difficult to make.

EVOLUTION OF INTERFACE RESIDUES

Residues that are of functional importance, such as those involved in enzyme active sites and ligand binding, are often conserved during evolution. However, the picture of conservation of residues involved in protein interfaces is less clear. One study concluded, from an analysis of five enzyme families, that interface residues were only slightly more conserved than the remainder of the protein sequence.[49] Another study of six homodimers concluded that interface residues were substantially more conserved than other surface residues and that this was particularly true of those residues that were completely buried in the central zone of the interface.[50] More recently an analysis of a much larger dataset of 64 protein interfaces also concluded that the protein interface is more conserved than the remainder of the protein surface.[51] However, this paper went on to conclude that when patches of surface residues where considered residue conservation was not sufficient to distinguish residues involved in protein interfaces from other surface residues. Rather, it was found that those surface patches with the highest conservation scores were parts of enzyme active sites.[51]

A study based on a much larger data set of 1494 protein associations (predominantly obligomers) has most recently concluded that for 63% of the associations the interface showed a higher level of conservation than the remainder of the protein surface. In addition for these interfaces it was possible to use conservation to discriminate interface sites.[16] They also conclude that transient interfaces also show the same levels of conservation as permanent interfaces.[16] Hence as more data has become there is agreement that interface sites have been conserved through evolution due to their functional importance.

CONCLUSION

A clearer understanding of the important factors that influence the formation of protein associations is beginning to emerge, as an increasing number of protein structures are being solved. The purpose of protein association is to achieve a structure that is energetically and functionally favourable. This requires the two components to present surfaces that are complementary in terms of physical and chemical characteristics. Complementarity is achieved through tight packing involving like-for-like secondary structure contacts. In addition the both hydrophobic effect and electrostatic effects make contributions to the energy of protein association. However the relative importance of these effects is still in question as different classes of protein association appear to exhibit different characteristics. Those associations that are permanent tend to have a greater contribution of hydrophobic interactions in their interfaces compared to transient associations, that have more polar interactions. Firmer conclusions can only be made when even larger data sets of classified protein associations become available.

As the size of the PDB[1] grows, so it will be possible to gather the larger datasets of protein associations necessary for comparative statistical analysis. However, the classification of protein associations in such large datasets presents a problem. Classification can only be achieved by integrating biochemical pathway and protein function information presented in journal articles. Hence, the automatic integration of functional information using novel text mining tools will be important for the future development of this field.

REFERENCES

1. Berman HM et al. The protein data bank. Nucleic Acids Res 2000; 28(1):235-242.
2. Jones S, Thornton JM. Principles of protein-protein interactions. Proc Natl Acad Sci U S A 1996; 93(1):13-20.
3. Nooren IMA, Thornton JM. Diversity of protein-protein interactions. EMBO J 2003; 22(14):3486-3492.
4. De S et al. Interaction preferences across protein-protein interfaces of obligatory and non-obligatory components are different. BMC Struct Biol 2005; 5.
5. Ansari S, Helms V. Statistical analysis of predominantly transient protein-protein interfaces. Proteins 2005; 61(2):344-355.
6. Ofran Y, Rost B. Analysing six types of protein-protein interfaces. J Mol Biol 2003; 325(2):377-387.
7. Ponstingl H, Kabir T, Thornton JM. Automatic inference of protein quaternary structure from crystals. J Appl Crystallogr 2003; 36:1116-1122.
8. Bahadur RP et al. A dissection of specific and nonspecific protein—Protein interfaces. J Mol Biol 2004; 336(4):943-955.
9. Henrick K, Thornton JM. PQS: a protein quaternary structure file server. Trends Biochem Sci 1998; 23(9):358-361.
10. Chothia C, Janin J. Principles of Protein-Protein Recognition. Nature 1975; 256(5520):705-708.
11. Janin J, Miller S, Chothia C. Surface, Subunit Interfaces and Interior of Oligomeric Proteins. J Mol Biol 1988; 204(1):155-164.

12. Miller S. The structure of interfaces between subunits of dimeric and tetrameric proteins. Protein Eng 1989; 3(2):77-83.
13. Lo Conte L, Chothia C, Janin J. The atomic structure of protein-protein recognition sites. J Mol Biol 1999; 285(5):2177-2198.
14. Jones S, Thornton JM, Protein-protein interactions—a review of protein dimer structures. Prog Biophys Mol Biol 1995; 63(1):13-20.
15. Ponstingl H et al. Morphological aspects of oligorneric protein structures. Prog Biophys Mol Biol 2005; 89(1):9-35.
16. Bordner AJ, Abagyan R. Statistical analysis and prediction of protein-protein interfaces. Proteins 2005; 60(3):353-366.
17. Bogan AA, Thorn KS. Anatomy of hot spots in protein interfaces. J Mol Biol 1998; 280(1):1-9.
18. Glaser F et al. Residue frequencies and pairing preferences at protein-protein interfaces. Proteins 2001; 43(2):89-102.
19. Chakrabarti P, Janin J. Dissecting protein-protein recognition sites. Proteins 2002; 47(3):334-343.
20. Chothia C. Nature of accessible and buried surfaces in proteins. J Mol Biol 1976; 105(1):1-14.
21. Jones S, Thornton JM. Analysis of protein-protein interaction sites using surface patches. J Mol Biol 1997; 272(1):121-132.
22. Sheinerman FB, Norel R, Honig B. Electrostatic aspects of protein-protein interactions. Curr Opin Struct Biol 2000; 10(2):153-159.
23. Bahar I, Jernigan RL. Inter-residue potentials in globular proteins and the dominance of highly specific hydrophilic interactions at close separation. J Mol Biol 1997; 266(1):195-214.
24. Zhou HX, Shan YB. Prediction of protein interaction sites from sequence profile and residue neighbor list. Proteins 2001; 44(3):336-343.
25. Nooren IMA, Thornton JM. Structural characterisation and functional significance of transient protein-protein interactions. J Mol Biol 2003; 325(5):991-1018.
26. Jiang SL, Tovchigrechko A, Vakser IA. The role of geometric complementarity in secondary structure packing: A systematic docking study. Protein Sci 2003; 12(8):1646-1651.
27. Rodier F et al. Hydration of protein-protein interfaces. Proteins 2005; 60(1):36-45.
28. Monecke P et al. Determination of the interfacial water content in protein-protein complexes from free energy simulations. Biophys J 2006; 90(3):841-850.
29. Janin J. Wet and dry interfaces: the role of solvent in protein-protein and protein-DNA recognition. Structure with Fold Des 1999; 7(12):R277-R279.
30. Hubbard SJ, Argos P. Cavities and packing at protein interfaces. Protein Sci 1994; 3(12):2194-2206.
31. Kauzmann W. Some factors in the interpretation of protein denaturation. Adv Protein Chem 1959; 14:1-63.
32. Dill KA. Dominant forces in protein folding. Biochemistry 1990; 29(31):7133-7155.
33. Xu D, Lin SL, Nussinov R. Protein binding versus protein folding: The role of hydrophilic bridges in protein associations. J Mol Biol 1997; 265(1):68-84.
34. Tsai CJ et al. Studies of protein-protein interfaces: A statistical analysis of the hydrophobic effect. Protein Sci 1997; 6(1):53-64.
35. Sundberg EJ et al. Estimation of the hydrophobic effect in an antigen-antibody protein-protein interface. Biochemistry 2000; 39(50):15375-15387.
36. Li YL et al. Magnitude of the hydrophobic effect at central versus peripheral sites in protein-protein interfaces. Structure 2005; 13(2):297-307.
37. Ratnaparkhi GS, Varadarajan R. Thermodynamic and structural studies of cavity formation in proteins suggest that loss of packing interactions rather than the hydrophobic effect dominates the observed energetics. Biochemistry 2000; 39(40):12365-12374.
38. Sheinerman FB, Honig B. On the role of electrostatic interactions in the design of protein-protein interfaces. J Mol Biol 2002; 318(1):161-177.
39. Froloff N, Windemuth A, Honig B. On the calculation of binding free energies using continuum methods: Application to MHC class I protein-peptide interactions. Protein Sci 1997; 6(6):1293-1301.
40. Schapira M, Totrov M, Abagyan R. Prediction of the binding energy for small molecules, peptides and proteins. J Mol Recognit 1999; 12(3):177-190.
41. Hendsch ZS, Tidor B. Do salt bridges stabilize proteins—a continuum electrostatic analysis. Protein Sci 1994; 3(2):211-226.
42. Kortemme T, Baker D. A simple physical model for binding energy hot spots in protein-protein complexes. Proc Natl Acad Sci U S A 2002; 99(22):14116-14121.
43. Janin J, Chothia C. The structure of protein-protein recognition sites. J Biol Chem 1990; 265(27):16027-16030.
44. Echols N, Milburn D, Gerstein M. MolMovDB: analysis and visualization of conformational change and structural flexibility. Nucleic Acids Res 2003; 31(1):478-482.
45. Betts MJ, Sternberg MJE. An analysis of conformational changes on protein-protein association: implications for predictive docking. Protein Eng 1999; 12(4):271-283.

46. Schneider TR. Objective comparison of protein structures: error-scaled difference distance matrices. Acta Crystallogr D Biol Crystallogr 2000; 56:714-721.
47. Ji ZL et al. KDBI: Kinetic Data of Bio-molecular Interactions Database. Nucleic Acids Res 2003; 31:255-257.
48. Kumar MDS, Gromiha MM. PINT: Protein-protein interactions thermodynamic database. Nucleic Acids Res 2006; 34:D195-D198.
49. Grishin NV, Phillips MA. The subunit interfaces of oligomeric enzymes are conserved to a similar extent to the overall protein sequences. Protein Sci 1994; 3(12):2455-2458.
50. Valdar WSJ, Thornton JM. Protein-protein interfaces: Analysis of amino acid conservation in homodimers. Proteins 2001; 42(1):108-124.
51. Caffrey DR et al. Are protein-protein interfaces more conserved in sequence than the rest of the protein surface? Protein Sci 2004; 13(1):190-202.
52. Rees B et al. Cardiotoxin-V4ii from Naja-Mossambica-Mossambica—the refined crystal-structure. J Mol Biol 1990; 214(1):281-297.
53. Lapthorn AJ et al. Crystal-structure of human chorionic-gonadotropin. Nature 1994; 369(6480):455-461.
54. Kresge N, Vacquier VD, Stout CD. 1.35 and 2.07 angstrom resolution structures of the red abalone sperm lysin monomer and dimer reveal features involved in receptor binding. Acta Crystallogr D Biol Crystallogr 2000; 56:34-41.
55. Kawashima T et al. The structure of the Escherichia coli EF-Tu center dot EF-Ts complex at 2.5 angstrom resolution. Nature 1996; 379(6565):511-518.

CHAPTER 4

DEATH BY CASPASE DIMERIZATION

Sarah H. MacKenzie and A. Clay Clark*
Department of Molecular and Structural Biochemistry, North Carolina State University, Raleigh, North Carolina, USA
**Corresponding Author: A. Clay Clark—Email: clay_clark@ncsu.edu*

Abstract: Controlled cell death, or apoptosis, occurs in response to many different environmental stimuli. The apoptotic cascade that occurs within the cell in response to these cues leads to morphological and biochemical changes that trigger the dismantling and packaging of the cell. Caspases are a family of cysteine-dependent aspartate-directed proteases that play an integral role in the cascade that leads to apoptosis. Caspases are grouped as either initiators or effectors of apoptosis, depending on where they enter the cell death process. Prior to activation, initiator caspases are present as monomers that must dimerize for full activation whereas effector caspases are present as dimeric zymogens that must be processed for full activation. The stability of the dimer may be due predominately to the interactions in the dimer interface as each caspase has unique properties in this region that lend to its specific mode of activation. Moreover, dimerization is responsible for active site formation because both monomers contribute residues that enable the formation of a fully functional active site. Overall, dimerization plays a key role in the ability of caspases to form fully functional proteases.

INTRODUCTION: APOPTOSIS LEADS TO CELL DEATH

Apoptosis is a type of cell death in which a cell uses specialized machinery to dismantle itself. Under normal growth and developmental conditions, apoptosis is a cell suicide mechanism that enables eumetazoans to control cell number, that is, to maintain homeostasis and to eliminate damaged cells. A healthy adult human produces approximately ten billion cells each day by mitosis and a similar number are removed by apoptosis.[1] Disregulation of the cell death mechanism results in a loss of homeostasis. Indeed, alterations in the cell death program have been implicated in several diseases,

Protein Dimerization and Oligomerization in Biology, edited by Jacqueline M. Matthews.

including neurodegenerative disorders, inflammatory diseases and cancer.[2] Cancer cells, in particular, are known to have a decreased sensitivity to proapoptotic signals when compared to normal cells. It is well established, however, that anticancer drugs are effective at inducing the cell death program by a variety of mechanisms.[3-5]

Apoptosis is a highly regulated process that may be triggered by a variety of stimuli including, but not limited to, virus infection, toxic stress, environmental insults and hormones.[6] The morphology of the cell changes during apoptosis due to cytoplasmic shrinkage, active membrane blebbing, chromatin condensation and fragmentation of membrane-enclosed vesicles.[7] In addition, the nuclear DNA is degraded, the cytoskeleton in dismantled and cell cycle progression is halted.[8,9] In short, every aspect of the cell is disrupted so that the contents are dismantled and packaged into vesicles, called apoptotic bodies, which are phagocytozed by macrophages or surrounding tissue.[10]

CASPASES ARE PART OF THE CELL DEATH MACHINERY

A family of cysteine-dependent aspartate-directed proteases, known as caspases, is intimately involved in apoptosis. The cleavage of key proteins in the cell by caspases leads to the morphological and biochemical changes observed in apoptosis. For example, the cleavage of ICAD (inhibitor of CAD) by caspases releases the DNase CAD (caspase activated DNase) from an inactive complex and ultimately results in the cleavage of nuclear DNA by CAD.[11] To date, fourteen caspases have been identified, with eleven caspases present in humans.[6,12]

Depending on their involvement in the life and death of a cell, caspases are broadly classified either as apoptotic or inflammatory caspases (Fig. 1). Those involved in the inflammatory response, namely caspases-1, -4 and -5, are cytokine activators.[13-15] The apoptotic caspases are further divided into two groups, the initiators and the effectors, depending on their time of entry into the apoptotic cascade (Fig. 1). Initiator caspases such as caspases-2, -8, -9 and -10 have an early entry into the cascade and are responsible for activating the effector caspases (-3, -6, or -7). They are themselves activated either by so-called extrinsic or intrinsic mechanisms.

The extrinsic pathway for initiator caspase activation ultimately is responsible for the elimination of unwanted cells that are produced during development or that have tumorogenic qualities.[16] This pathway is initiated by ligation of a transmembrane death receptor in response to an extracellular signal, followed by recruitment and activation of initiator caspases as a part of a multiprotein complex (Fig. 2). Caspases-8 and -10 are the initiator caspases that are activated by way of the extrinsic pathway. In contrast, the intrinsic pathway primarily is responsible for the removal of cells in response to cytotoxic stress, chemotherapeutic drugs, mitochondrial damage and certain developmental cues.[17] The mitochondria release cytochrome c into the cytoplasm in response to one or more of these cues (Fig. 2). The increase in the cytoplasmic concentration of cytochrome c is sensed by the protein Apaf-1 (apoptosis activating factor 1), which leads to recruitment of caspase-9 to a multiprotein complex, called the apoptosome, followed by activation of the caspase in a cofactor-dependent manner. The end result of initiator caspase activation is the downstream activation of the effector caspases-3, -6 and -7, which ultimately are responsible for cleavage of intracellular proteins that lead to the dismantling of the cell.

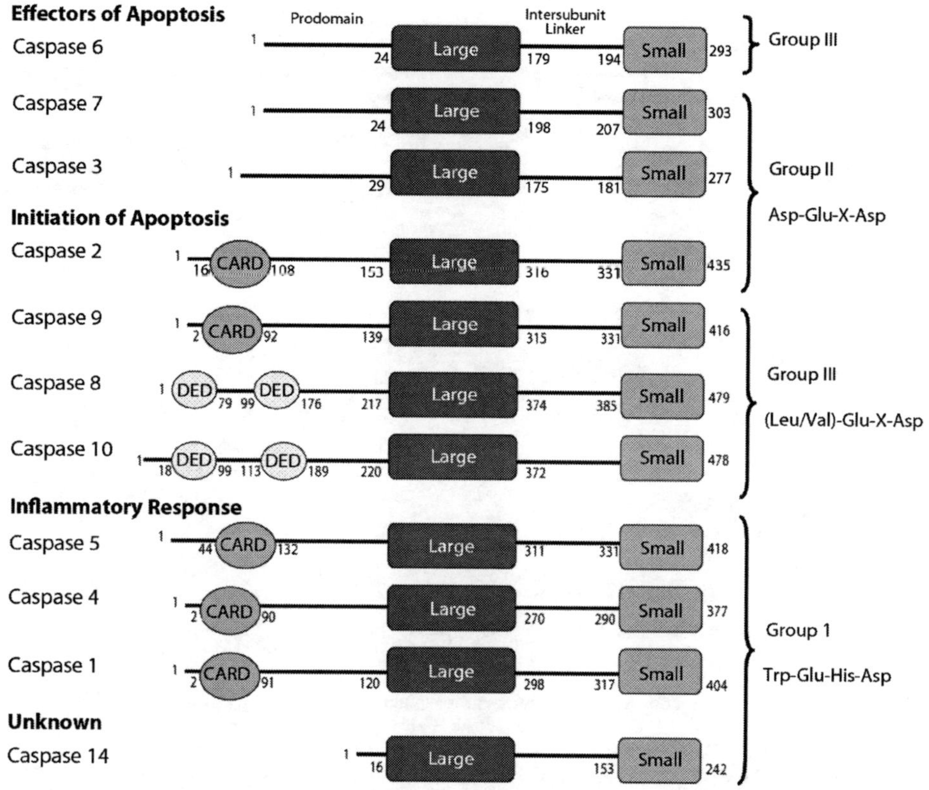

Figure 1. Human caspase organization. Caspases are grouped on the left according to function and on the right according to the recognition sequence of the substrate. Each caspase has an N-terminal prodomain, where some contain either a CARD (caspase recruitment domain) or DED (death effector domain) motif, followed by the large subunit (LARGE), an intersubunit linker and the small subunit (SMALL). The numbers on each caspase molecule refers to the length of each specific domain, which was determined using the NCBI domain organization database (http://www.ncbi.nlm.nih.gov/).

CASPASES ALSO ARE PART OF THE INFLAMMATORY RESPONSE

Some members of the caspase family are involved in processes within the cell that are not lethal. Caspases-1, -4 and -5 function in the regulation of inflammatory processes, for example. Caspase-1, which was known originally as interleukin-1β converting enzyme (ICE), is responsible for the activation of interleukin-1β (IL-1β) in macrophages[13] as well as interleukin-18 (IL-18), a member of the IL-1 cytokine superfamily.[18] Caspase-5 interacts with caspase-1 in a multiprotein complex known as the inflammasome, where caspase-1 is activated[19] (Fig. 2). The function of caspase-4 is more obscure, although it has been shown that its transcription is induced by interferons.[15]

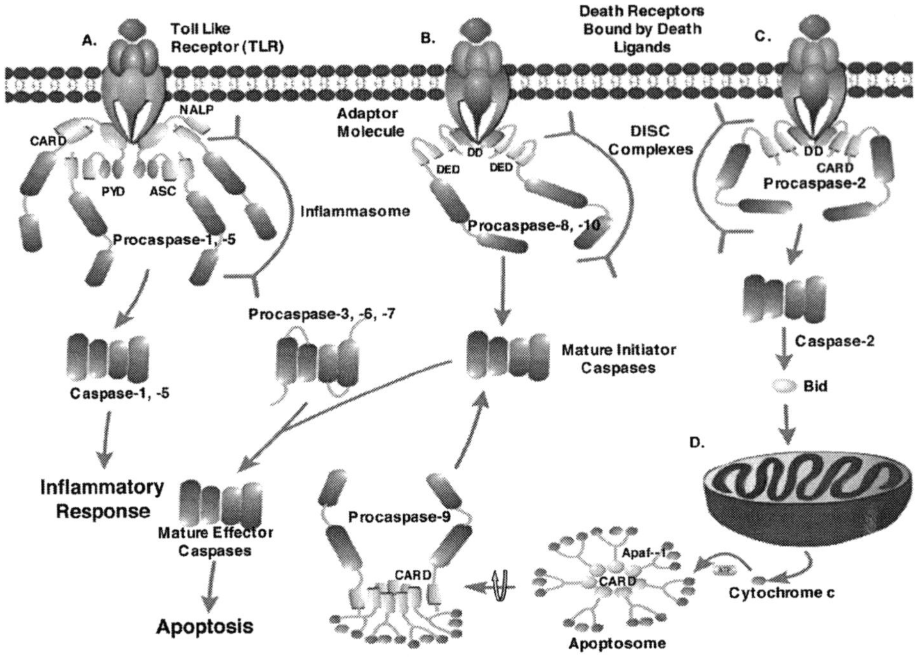

Figure 2. The caspase cascade. A) Inflammatory caspase activation: a ligand binds to a toll-like receptor (TLR), which signals a NALP protein to bind to the TLR. ASC interacts with the pyrin domain (PYD) of NALP via PYD:PYD interactions. The CARD domain of ASC interacts with the CARD domain of procaspase-1 and the CARD domain of NALP interacts with the CARD domain of procaspase-5, forming the inflammasome. The inflammasome complex promotes dimerization of caspases-1 and -5, leading to their activation and the inflammatory response. B) The extrinsic apoptotic pathway: a death ligand binds to a death receptor, which signals an adaptor molecule to bind to the receptor via death domain (DD) interactions. The DED motif of the adaptor molecule interacts with the DED of procaspases-8 and -10, forming a DISC complex. Dimerization (mechanism unknown) results in maturation and full activity. Caspases-8 and -10 then process executioner caspases. C) Procaspase-2, a unique caspase, is activated when a ligand binds to a death receptor, which signals an adaptor molecule to bind via interactions with the death domain. The CARD of the adaptor molecule interacts with the CARD of procaspase-2 to promote dimerization in a DISC-like complex. Upon removal of the prodomain, caspase-2 cleaves Bid, a protein responsible for the increased permeability of the mitochondria. D) The intrinsic apoptotic pathway: an increase in the cytosolic concentration of cytochrome c leads to the formation of the apoptosome. The apoptosome is composed of Apaf-1 monomers that form a heptameric structure when cytochrome c binds to the WD40 motifs of Apaf-1, in an ATP-dependent manner, leading to interactions of the CARDs. The CARD of procaspase-9 then interacts with the CARD of Apaf-1, increasing the local concentration of procaspase-9 monomers and thereby promoting dimerization and activation. Caspase-9 then processes effector caspases, which leads to apoptosis. Effector caspases are activated by cleavage of their prodomain and intersubunit linker.

CASPASES CLEAVE WITH HIGH SPECIFICITY

Caspases show a high degree of specificity for substrates, with an absolute requirement for an aspartic acid residue and a recognition sequence of four amino acids N-terminal to the cleavage site. These residues generally are referred to as positions 1-4, or P1-P4, where the aspartate occupies the P1 position and cleavage of the substrate occurs C-terminal to the P1 aspartate. In addition to the functional groupings described above, caspases also have been grouped into three sub-families based on consensus recognition sequences.[20] Group I caspases (caspases-1, -4, -5 and -14) recognize Trp-Glu-His-Asp; Group II caspases (caspases-2, -3 and -7) recognize Asp-Glu-X-Asp, where X is any amino acid; and Group III (caspases-6, -8, -9 and -10) recognize (Leu/Val)-Glu-X-Asp (Fig. 1). The specificity of the sub-families is crucial to the apoptotic mechanism because it prevents indiscriminate proteolysis of proteins that are not destined to be destroyed in the cell death process.[21]

CASPASES ARE PRODUCED INITIALLY AS INACTIVE ZYMOGENS

Many of the events that occur during apoptosis are posttranslational in nature. Caspases are synthesized initially as zymogens (procaspases) and the pool of inactive zymogen is converted rapidly to active protease upon induction of apoptosis. While some procaspases are monomers and some are dimers, as described below, all procaspases have a common organization in that each monomeric unit is composed of an N-terminal prodomain, a large subunit and a small subunit, as shown schematically in Figure 1. In all procaspases, the large and small subunits are covalently connected by a sequence of amino acids referred to as the intersubunit linker. The linker is cleaved and in some cases completely excised, upon maturation. At the N-terminus, the caspase prodomain ranges in length from 15 amino acids for caspase-14 to 219 amino acids for caspase-10. In general, initiator caspases contain a long prodomain and effector caspases contain a short prodomain.

The prodomains of caspases have been shown to have a variety of functions, from facilitating dimerization,[22-24] to assisting in the folding of the caspase as an intramolecular chaperone,[25] to sequestering the protein in the cytoplasm,[26] to silencing the zymogen in vivo.[27,28] The long prodomain of initiator caspases contains recognition motifs that are involved in homo- or heterotypic interactions that lead to zymogen activation. For example, the death effector domain (DED) is found within the prodomain of caspases-8 and -10 and interacts with an adaptor molecule that is recruited upon ligation of a death receptor, as shown schematically in Figure 2. Utilizing a similar activation protocol, the caspase recruitment domain (CARD) is found within the prodomain of caspases-1, -2, -4, -5 and -9 and is involved in recognition of the caspase by the particular multiprotein complex that facilitates activation.

An active caspase molecule is derived from the processing and association of two procaspase monomers. Structurally, mature caspases are dimers of heterodimers that contain two copies each of the large and small subunits. The dimer is arranged in an LSSL configuration, where L represents the large subunit and S represents the small subunit, with two active sites at nearly opposite ends of the molecule. The dimer is composed of a 12-stranded β-sheet core surrounded by a network of α-helices (Fig. 3A) and the substrate-binding region is formed on the protein surface by five loops (L1, L2, L3, L4 and L2'). L1-L4 are contributed by one heterodimer and L2' is contributed by the second heterodimer.

DIMERIZATION OF CASPASE ZYMOGENS

For all caspases, dimerization is critical for proper active site formation. Since the active site loops are contributed by both heterodimers in the dimeric structure (Fig. 3), caspase zymogens demonstrate little or no activity in the monomeric state. Overall, the initiator

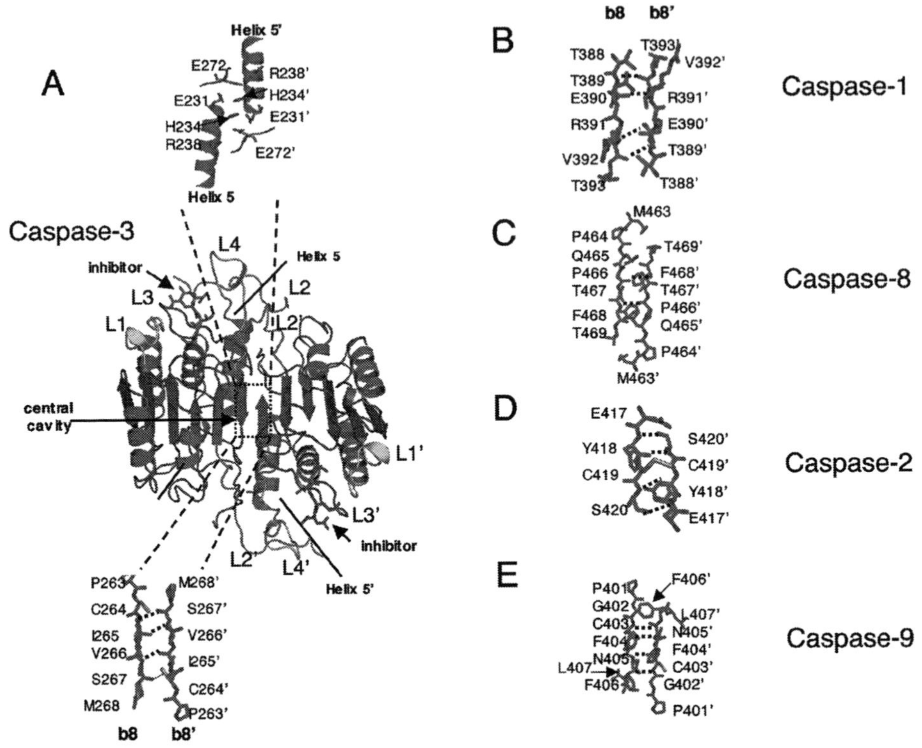

Figure 3. Comparison of caspase dimer interfaces. A) Structure of caspase-3 (PDB entry 2J30). α-helices 5 and 5' and active site loops L1, L2, L2', L3 and L4 are labeled. The prime (') indicates residues from the second heterodimer. Residues in β-strands 8 and 8' are shown below the structure and ionic interactions between residues in helices 5 and 5' are shown above the structure rotated by 180° to view sidechains. Amino acids in the dimer interface (β-strands 8 and 8') of caspase-1 (B) (PDB entry 2HBQ), caspase-8 (C) (PDB entry 1QTN), caspase-2 (D) (PDB entry 1PYO) and of caspase-9 (E) (PDB entry 1JXQ) are shown. For A-E, the dashed lines indicate main chain hydrogen bonds between β-strands 8 and 8'. Structures were generated using *Pymol* (Delano Scientific LLC, Palo Alto, CA).

procaspases are thought to dimerize by an induced proximity model where oligomeric activation complexes mediate clustering of zymogens and increase the local concentration of procaspase monomers, which in turn facilitates dimerization.[29,30] For example, procaspases-8, -9 and -10, the initiator caspases, are found as monomers at physiological concentrations and must undergo dimerization to become active. Once an extracellular ligand binds to the death receptor, the intracellular region of the receptor binds to the death domain (DD) of an adaptor molecule (Fig. 2). The adaptor molecule then recruits procaspase-8 or -10 by interacting with the death effector domain (DED) located in the long prodomain. The resulting assembly is known as the death-inducing signaling complex (DISC) (Fig. 2).[31] Likewise, Apaf-1 assembles into a heptameric structure in the presence of cytochrome c and ATP to recruit procaspase-9 through CARD-CARD interactions. The resulting assembly is referred to as the apoptosome (Fig. 2). For initiator caspases, the intersubunit linker is cleaved following dimerization and while this stabilizes the dimer, cleavage of the linker is not necessary for the gain of enzymatic activity since a functional active site can form without this event.[32]

A unique initiator caspase, caspase-2, is found as a dimer at physiological concentrations and is activated by proteolytic cleavage. This caspase functions as an initiator of both the intrinsic and extrinsic pathways and in some cases, as in neuronal cells, it acts as an effector caspase.[33,34] In the extrinsic pathway, caspase-2 initiates apoptosis in response to ligation of a death receptor utilizing the CARD of caspase-2. In a DISC-like complex, caspase-2 is proteolytically cleaved in the intersubunit linker and within the prodomain, which releases the active protease into the cytoplasm. Unlike other initiator caspases, caspase-2 is completely inactive toward the effector caspase zymogens. Caspase-2, instead, cleaves Bid (BH3 interacting death agonist), which leads to an increase in the permeability of the outer membrane of the mitochondria. This event elevates the cytosolic levels of cytochrome c, resulting in the formation of the apoptosome and the subsequent activation of caspase-9. The mechanism of caspase-2 activation in the intrinsic pathway remains obscure; however, it is known that caspase-2 plays an important role in initiating chemotherapy-induced cell death.[21]

Procaspases-1, -4 and -5, those involved in the regulation of inflammatory processes, are zymogens that exist in the cell as monomers and require dimerization to become active. Procaspases-1 and -5 can be activated by several different inflammasomes that contain a variety of proteins such as Ipaf, NALP1, NALP2, NALP3 and ASC. The inflammasomes are formed similarly to the multi-protein complexes of the extrinsic apoptotic pathway. That is, an interaction with a ligand and its cognate receptor results in the recruitment of an adaptor molecule that in turn binds the CARD of the inflammatory caspases (Fig. 2). The induced proximity effect, described above, apparently is utilized by these procaspases as well. The binding of the procaspases to the inflammasome increases the local concentration of monomers and thereby facilitates oligomerization.[35]

CASPASE DIMERIZATION AFFECTS STABILITY

In contrast to the initiator caspases, procaspases-3, -6 and -7, the effectors, are found as stable, but inactive, dimers at physiological concentration. As described below, the procaspases are inactive due to misaligned active sites that prevent efficient catalysis. Activation of these caspases occurs after proteolytic cleavage of the intersubunit linker, which allows the active site loops to rearrange.

The stability of caspases is due largely to dimerization. Studies of procaspase-3 showed that the protein unfolds by an unusual mechanism that includes two partially folded intermediates in equilibrium with the native dimer and the unfolded monomer.[36,37] One intermediate is an inactive dimer, the second is a partially folded monomer. An examination of the free energy change for the formation of each species in the unfolding process showed that dimerization had a significant contribution to the conformational free energy of the protein. While the monomer has a conformational free energy of ~7 kcal/mol (25°C and pH 7.2), the protein gains an additional ~18 kcal/mol upon dimerization and subsequent isomerization of the inactive dimer.[36] Dimerization, therefore, results in a very stable protein under physiological conditions. Indeed, the equilibrium dissociation constant of the procaspase-3 dimer is estimated to be in the low nanomolar range or lower.[36]

Dimerization of procaspase-3 is a pH-dependent event. While procaspase-3 is a dimer at neutral pH, the dimer is destabilized in favor of the monomer at pH 4. Titration studies have shown that two events occur as the pH is lowered.[37] First, binding of the pro-domain to the protease domain is reduced between pH 7 and 5, although the protein remains dimeric. Second, the dimer dissociates between pH ~6 and 4. The dimeric state of procaspase-3 is unlikely to be affected during apoptosis because the pH of a cell decreases from 7.4 to ~6.8.[38] This may not be true for the mature caspase-3, however, since the dimer dissociates between pH ~7 and ~5.[39] Small fluctuations around pH 6.8 are predicted to have larger effects on the dimeric structure of caspase-3 than on that of procaspase-3. In contrast to procaspase-3, folding studies of monomeric caspases have not been done, so currently it is not known whether these principles hold true for other caspases.

A COMPARISON OF CASPASE DIMER INTERFACES

As described above, mature caspases are dimers of heterodimers, i.e., each heterodimer comprises a newly cleaved large and small subunit. The caspase structure includes a water-filled central cavity on one side and α-helices on the other side. Interactions across the dimer interface among charged amino acids in the α-helices are predicted to stabilize the structure (Fig. 3A). Approximately 2000 Å is buried in the dimer interface.[40,41]

Most of the interface interactions occur in β-strands 8 and 8' in the small subunits since they form the lining of the central cavity (Fig. 3A). In general, for initiator procaspases, the dimerization rather than the processing of the zymogen is the key event in activation. That is, the acquisition of enzymatic activity is coupled to dimer formation rather than to chain cleavage. This mechanism is fundamentally different in the effector procaspases since those proteins are stable dimers that have low activity.

An understanding of the molecular interactions at the dimer interface may help to explain why some caspase zymogens are monomers in the cell while others form stable dimers under physiological conditions. Although the three-dimensional structures are quite similar among the caspases, the dimer interfaces are very different. The network of hydrogen bonds, the presence or absence of salt bridges or other polar interactions and the hydrophobic nature of the dimer interface likely influence the stability of the dimer. The combination of these features could explain why caspase-8, for instance, remains associated with the DISC once activated. The multi-protein complex could help to stabilize the dimeric form of caspase-8 since the monomer appears to be the preferred structure in solution. The same scenario holds true for caspase-9. It remains associated with the apoptosome to carry out catalytic functions apparently because of the weak interactions at the dimer interface. On the other hand, caspase-3 forms a stable dimer in solution and thus does not require activation complexes to facilitate dimerization.

In addition to stabilizing the protein, dimerization also is critical to active site formation. All caspases contain five loops that form the active site. Four loops (L1, L2, L3 and L4) are contributed by one heterodimer and the fifth loop, L2', is contributed by the second heterodimer. As described below, the loops rearrange upon maturation of the protein, enabling the substrate to bind productively in the active site. The resulting conformation is stabilized by interactions in the dimer interface, although the specific stabilizing interactions are unique to each caspase as follows.

Caspase-3 contains four buried hydrogen bonds that are contributed by main chain atoms in β-strands 8 and 8' and span the dimer interface (see Fig. 3A). In addition, residues from helices 5 and 5' (Glu231, His234, Arg238 and Glu272, see Fig. 3A) create a network of salt-bridges on the surface of the protein that are predicted to stabilize the dimer. The region of the caspase-3 interface composed of β-strands 8 and 8' has, in general, a hydrophobic character, possibly explaining why procaspase-3 is found as a stable dimer in the cell. That is, the protein forms a stable dimer as a mechanism to shield the amino acids from solvent. Otherwise, the hydrophobic β-strand 8 would be solvent-exposed in the monomer.

In contrast, the dimer interface of *caspase-1* is more hydrophilic and presents a different hydrogen bonding network. The addition of Glu390 and Arg391 on β-strand 8 and Glu390' and Arg391' on β-strand 8' provides four additional charged residues in the center of the dimer (Fig. 3B). Arg391 forms a β-bulge that perturbs the pattern of hydrogen bonds in β-strand 8 and results in a so-called "negative design element".[42] The element prevents the formation of indiscriminate oligomers from an exposed β-strand by requiring a precise hydrogen bonding pattern. Overall, this is consistent with the presence of a procaspase-1 monomer in solution because the exposed β-strand 8 in the procaspase-1 monomer would hydrogen bond only with another monomer that presents a complementary surface. The side chain of Glu390 hydrogen bonds with the amide backbone nitrogen of Arg391' across the dimer interface and also forms a salt-bridge with Arg286, from the active site of the same heterodimer. This salt bridge is predicted to stabilize active site loop 2 (see also Fig. 5).[43]

The dimer interface of *caspase-8* contains only two main-chain hydrogen bonds, between Phe468 and Pro466' as well as the symmetry-related Phe468' and Pro466 interaction (Fig. 3C). The interface of caspase-8 is hydrophilic in nature and is striking for the absence of an arginine residue next to the catalytic cysteine, C360. Glutamine occupies this position in caspase-8 (Table 1). Currently, it is not known how the polar interface might contribute to the formation of a stable monomer rather than a stable dimer in solution. Current models suggest that the protein-protein interactions between the pro-domain and the DISC result in dimerization of caspase-8 by increasing the local concentration of interacting species, although the mechanistic details of initiator caspase dimerization are not known.

Caspase-2 is unique among the caspases because it has a covalent linkage that spans the dimer interface. The central cysteine pair, Cys419 and Cys419', forms a disulfide bridge (Fig 3D and Table 1).[21] Interestingly, the residue at this position in many caspases is implicated in interactions across the dimer interface, such as Glu390 described above for caspase-1. However, the disulfide linkage is not required for dimerization of caspase-2,[34] so the role of the disulfide bond is not known at present.

The interface of *caspase-9* has three primary differences at the dimer interface that are distinct from those of other caspases, resulting in the formation of only one functional active site. First, caspase-9 contains a phenylalanine residue at position 404 (Fig. 3D and Table 1) rather than valine, glutamate, or cysteine at this position in caspases-3, -1, or -2, respectively (Fig. 3A, 3B and 3D). As a result, caspase-9 cannot simultaneous form two productive active sites due to steric clashes in the dimer interface. Second, a seven-residue loop is inserted between strands 3 and 4 of the large subunit (see Fig. 6A). Several residues in the insertion fill the central cavity on the surface of the protein and form polar contacts across the dimer interface.[32] Third, the intersubunit linker is about 20 residues longer than that of procaspase-3. Together, these features somehow allow for a zymogen that has a relatively high activity when compared to that of the activated caspase.

Table 1. Caspase sequence alignment

Caspase	8	1	MDFSRNLYDI	GEQLDSEDLA	SLKFLSLDYI	PQRKQEPIKD	41	ALMLFQRLQE	KRMLEESNLS	FLKELLFRIN	RLDLLI.TYL	80	NTRKEEMERE	LQTPGRAQIS
Caspase	1							1 MADK	VLKEKRKLFI	RSMGEGTING	LLDELL.QTR	34	VLNKEEMEKV	KRENATVMDK
Caspase	2							1 M	AADRGRRILG	VCGMHPHHQE	TLKKNRVVLA	32	KQLLLSELLE	HLLEKDIITL
Caspase	9							1	MDEADRRLLR	RCRLRLVEEL	QVDQLWDALL	31	SSELFRPHMI	EDIQRAGSGS
Caspase	8	100	AYRVMLYQIS	EEVSRSELRS	FKFLLQEEIS	KCKLDDDMNL	140	LDIFIEMEKR	VILGEGKLDI	LKRVCAQINK	SLLKIINDYE	180	EFSKERSSSL	EGSPDEFSNG
Caspase	1	54	TRALIDSVIP	KGAQACQICI	TYICE.....	....	79	....	....	....	...EDSYLAG	86	TLGLSADQTS	GNYLNMQDSQ
Caspase	2	52	EMRELIQAKV	GSFSQNVELL	NLLPKRGPQA	FDAFCEALRE	92	....	....	......TKQG	HLEDMLLTTL	106	SGLQHVLPPL	SCDYDLSLPF
Caspase	9	51	RRDQARQLII	DLETRGSQAL	PLFISCLEDT	GQDMLASFL.	90	....	....	.....RTNRQ	AGKLSKPTLE	105	NLTPVVLRPE	IRKPEVLRPE
Caspase	3											1		MENTENSVDS
Caspase	7											1		MADDQGCIEE
Caspase	8	200	EELCGVMTIS	DSPREQDSES	QTLDKVYQM.	....	229	....	....KSKPRG	YCLIINNHNF	AKAREKVPKL	255	HSIRDRNGTH	LDAGALTTTF
Caspase	1	106	GVLSSFPAPQ	AVQDNPAMPT	SSGSEGNVKL	CSLEEA..QR	144	IWKQKSAEIY	PIMDKSSRTR	LALIICNEEF	....	174	DSIPRRTGAE	VDITGMTMLL
Caspase	2	126	PVCESCPLYK	KLRLSTDTVE	HSLDNKDGPL	CLQVKPCTPE	166	FYQTHFQLAY	RL..QSRPRG	LALVLSNVHF	.......TGE	197	KELEFRSGGD	VDHSTLVTLF
Caspase	9	125	TPRP......	....	.....VDIGS	GGFGDVGALE	144	SLRGNADLAY	IL..SMEPCG	HCLIINNVNF	.......CRE	175	SGLRTRTGSN	IDCEKLRRRF
Caspase	3	11	KSIKNLEPKI	IHGSESMDSG	ISLDNSYKM.	....	40	....	....DYPEMG	LCIIINNKNF	.......HKS	59	TGMTSRSGTD	VDAANLRETF
Caspase	7	11	QGVEDSANED	SVDAKPDRSS	FVPSLFSKKK	KNVTMRSIKT	51	TRDRVPTYQY	NM..NFEKLG	KCIIINNKNF	.......DKV	82	TGMGVRNGTD	KDAEALFKCF
Caspase	8	275	EELHFEIKPH	DDCTVEQIYE	ILKIYQL.MD	HSNMD..CFI	312	CCILSHGDKG	IIYGTD....	...GQEPPIY	ELTSQFTGLK	345	CPSLAGKPKV	FFIQACQGDN
Caspase	1	194	QNLGYSVDVK	KNLTASDMTT	ELEAFAHRPE	HKTSD..STF	232	LVFMSHGIRE	GICGKK..HS	EQVPDILQLN	AIFNMLNTKN	270	CPSLKDKPKV	IIIQACRGDS
Caspase	2	217	KLLGYDVHVL	CDQTAQEMQE	KLQNFAQLPA	HRVTD..SCI	255	VALLSHGVEG	AIYGVD....	...GKLLQLQ	EVFQLFDNAN	288	CPSLQNKPKM	FFIQACRGDE
Caspase	9	195	SSLHFMVEVK	GDLTAKKMVL	ALLELAR.QD	HGALD..CCV	232	VVILSHGCQA	SHLQFPGAVY	GTDGCPVSVE	KIVNIFNGTS	272	CPSLGGKPKL	FFIQACGGEQ
Caspase	3	79	RNLKYEVRNK	NDLTREEIVE	LMRDVSK.ED	HSKRS..SFV	116	CVLLSHGEEG	IIFGTN....	....GPVDLK	KITNFFRGDR	148	CRSLTGKPKL	FIIQACRGTE
Caspase	7	102	RSLGFDVIVY	NDCSCAKMQD	LLKKASE.ED	HTNAA..CFA	139	CILLSHGEEN	VIYGKD....	....GVTPIK	DLTAHFRGDR	171	CKTLLEKPKL	FFIQACRGTE

continued on next page

Table 1. Continued

Caspase	8	365	YQKGIPVETD	SEEQ......	.........P	YLEMDLSSPQ	390	TRYIPDEA..	..DFLLGMAT	VNNCVSYRNP	AEGTWYIQSL	426	CQSLRERCPR	GDDILTILTE
Caspase	1	290	PGVVW..FKD	SVGVSGN...	....LSLPT	TEEFEDDAIK	320	KAHIEK....	..DFIAFCSS	TPDNVSWRHP	TMGSVFIGRL	354	IEHMQEYACS	.CDVEEIFRK
Caspase	2	308	TDRGVD.QQD	GKNHAG....	........S	PGCEESDAGK	334	EKLPKMRLPT	RSDMICGYAC	LKGTAAMRNT	KRGSWYIEAL	374	AQVFSERACD	.MHVADMLVK
Caspase	9	292	KDHGFEVAST	SPEDESPGSN	PEPDATPFQE	GLRTFDQLDA	332	ISSLPTPS..	..DIFVSYST	FPGFVSWRDP	KSGSWYVETL	368	DDIFEQWAHS	.EDLQSLLLR
Caspase	3	168	LDCGI..ETD	SGVD......	....	......DDMA	184	CHKIPVEA..	..DFLYAYST	APGYYSWRNS	KDGSWFIQSL	220	CAMLKQYADK	.LEFMHILTR
Caspase	7	191	LDDGI..QAD	SGPI......	....	...NDTDANP	210	RYKIPVEA..	..DFLFAYST	VPGYYSWRSP	GRGSWFVQAL	246	CSILEEHGKD	.LEIMQILTR
Caspase	8	446	VNYEVSNKDD	......KKNM	GKQMPQPTFT	LRKKLVFPSD								
Caspase	1	373	VRFSFEQPD.	.........G	FAQMPTTERV	TLTRCFYLFP	403	GH						
Caspase	2	393	VNALIKDREG	YAPGTEFHRC	KEMSEYCSTL	CRHLYLFPGH	433	PPT						
Caspase	9	387	VANAVSVKG.	I	YKQMPGCFNF	LRKKLFFKTS								
Caspase	3	239	VNRKVATEFE	SFSFDATFHA	KKQIPCIVSM	LTKELYFYH								
Caspase	7	265	VNDRVARHFE	SQSDDPHFHE	KKQIPCVVSM	LTKELYFSQ								

Sequence data were obtained using NCBI (http://www.ncbi.nlm.nih.gov/)

ACTIVE SITE FORMATION AFTER CHAIN CLEAVAGE

Structural studies of caspase-7,[45,46] an effector caspase and of caspases-1[47,48] and -9[32] provide clues to conformational changes that occur during maturation. In general, the active site is misaligned in the zymogen and cleavage of the intersubunit linker allows for realignment of the active site loops. While the mechanistic details differ for caspases-7 and -1, both proteins follow the general scheme that the chain cleavage results in proper formation of the substrate binding pocket as well as formation of the loop bundle among L2, L4 and L2' (Fig. 4A). Formation of the loop bundle stabilizes the active site configuration and moves the catalytic cysteine into the S1 subsite where it is positioned to attack the substrate at the P1 site.

For caspase-7, cleavage of the intersubunit linker occurs at Asp198. Prior to the cleavage event, one of the two intersubunit linkers binds in the central cavity whereas the other is positioned on the outside of the cavity (Fig. 4B). Following cleavage, residues in active site loop 3 move into the central cavity and toward the dimer interface, resulting in formation of the substrate-binding groove (Fig. 4C). This insertion is prevented in the zymogen by the blocking segment of the intersubunit linker. More specifically, amino acids 212-215 of one monomer occupy this position in the central cavity and prevent movement of the so-called elbow loop of the second monomer into the central cavity

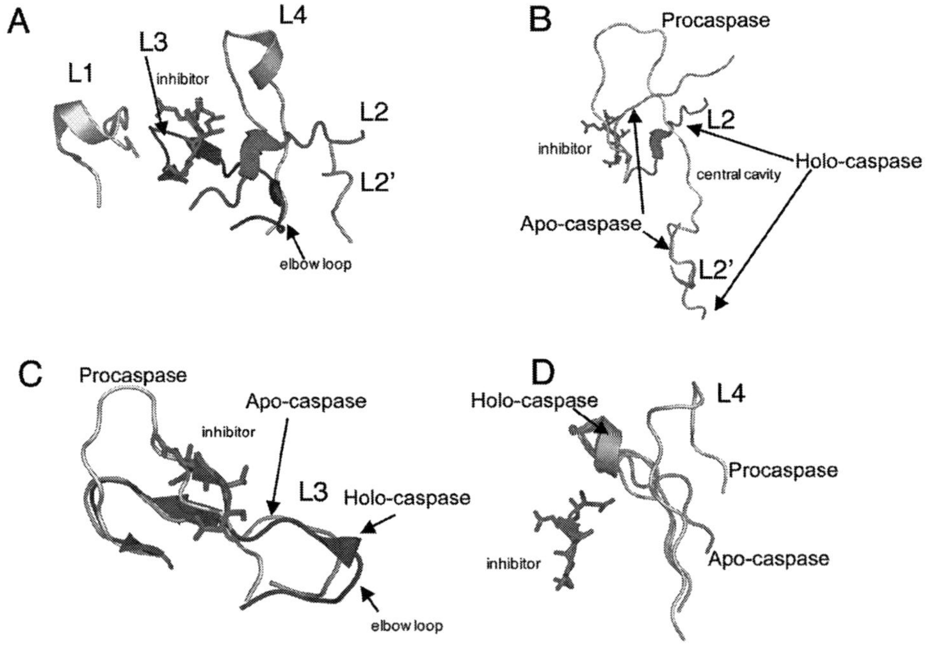

Figure 4. Active site loop movements upon maturation of caspase-7. A) Active site loops 1-4 and 2'. B) Movements in L2 and L2' upon maturation and substrate binding. Holo-caspase-7 (PDB entry 1F1J), apo-caspase-7 (PDB entry 1K86), procaspase-7 (PDB entry 1GQF). Note that in procaspase-7 the intersubunit linker encompasses both L2 and L2'. C) Movements in L3 upon maturation. D) Movements in L4 upon maturation. For A-D, the inhibitor bound to holo-caspase-7 is indicated. Structures were generated using *Pymol* (Delano Scientific LLC, Palo Alto, CA).

(Fig. 4C). The elbow loop contains residues 224-231 and, in the activated caspase, forms several backbone hydrogen bonds with L2 as well as stacking interactions with other side chains. Thus, preventing insertion of the elbow loop destabilizes the substrate-binding pocket as well as the catalytic loop (L2). As a result, L3 in the procaspase occupies the substrate-binding pocket (Fig. 4C). Proteolytic cleavage of the intersubunit linker allows the formation of a new N-terminus of the small subunit (L2') and C-terminus of the large subunit (L2). L2' leaves the central cavity and rotates by ~180° to form new contacts in the loop bundle with L2, L3 and L4 from the opposing monomer (Fig. 4A,B).

The structure of apo-caspase-7, determined in the absence of inhibitor, reveals an intermediate form in which the active site is neither fully active nor fully misaligned. For example, one observes that L2' remains bound in the central cavity and retains the misaligned conformation (Fig. 4B); that is, it resembles the configuration of the zymogen rather than that of the holo-enzyme. As a consequence, the loop bundle (L2, L4, L2') is not properly assembled. In contrast, L3 (Fig. 4C) and L4 (Fig. 4D) are positioned closer to their productive conformations, although they are not fully stabilized until L2' provides the additional interactions. Once L2' rotates out of the central cavity upon substrate binding, the elbow loops from each heterodimer can insert into the cavity and a functional active site forms. More specifically, the arginine positioned next to the catalytic cysteine rotates from a solvent exposed orientation in the substrate binding pocket toward the dimer interface and intercalates between tyrosine 223 on β-strand 7 and proline 227 on the elbow loop, to form an ordered lining to the central cavity (Fig. 5D). The reorientation of the arginine on L2 toward the dimer interface and the accompanying loop movements lock the S1-S4 subsites into place and position the backbone amide of the catalytic cysteine into the correct orientation to form part of the oxyanion hole.[45,46] These findings suggest that substrate binding plays a key role in the full activation of caspases and that proteolytic cleavage of the intersubunit linker is not sufficient for activation. The caspase active site is dynamic in that it fluctuates between the misaligned conformation, in which L2' occupies the central cavity and the substrate binding pocket is not formed and the active conformation, in which formation of the loop bundle allows the active site to align fully. The binding of substrate traps the protein in the active conformation.

The formation of the active site of caspase-1 is somewhat different than that of caspase-7. As described above, β-strand 8 contains a bulge in the region that accommodates the blocking segment in (pro)caspase-7 due to the insertion of Arg391. In order for the formation of the active site of caspase-1 to mimic that of caspase-7, Arg391 would have to interact with another residue so that the blocking segment can fit into the central cavity. Structural data for caspase-1 reveal that in the ligand-free form the active site loops exhibit a misaligned conformation that is catalytically inactive prior to substrate binding.[48] Three segments (L2, L3 and L4) undergo conformational changes upon substrate/inhibitor binding, resulting in catalytically active caspase-1. In particular, Arg341 in L3 rotates from a solvent exposed position on the surface of the enzyme into the S1 pocket of the active site (Fig. 5A). Upon substrate binding, Arg341 forms hydrogen bonds with amino acids in L1 and residues in the P1 and P3 positions of the substrate. The movement of Arg341 creates movement of the entire loop to place other key residues in their catalytically active form (e.g., Trp340 rotates from the surface of the protein into the S2 region of the active site, where it hydrogen bonds with the substrate; Fig. 5A).

The ligand-free form of caspase-1 lacks the salt bridge between Arg286 and Glu390 that stabilizes the active site and the dimer interface in the ligand-bound form (Fig. 5C).

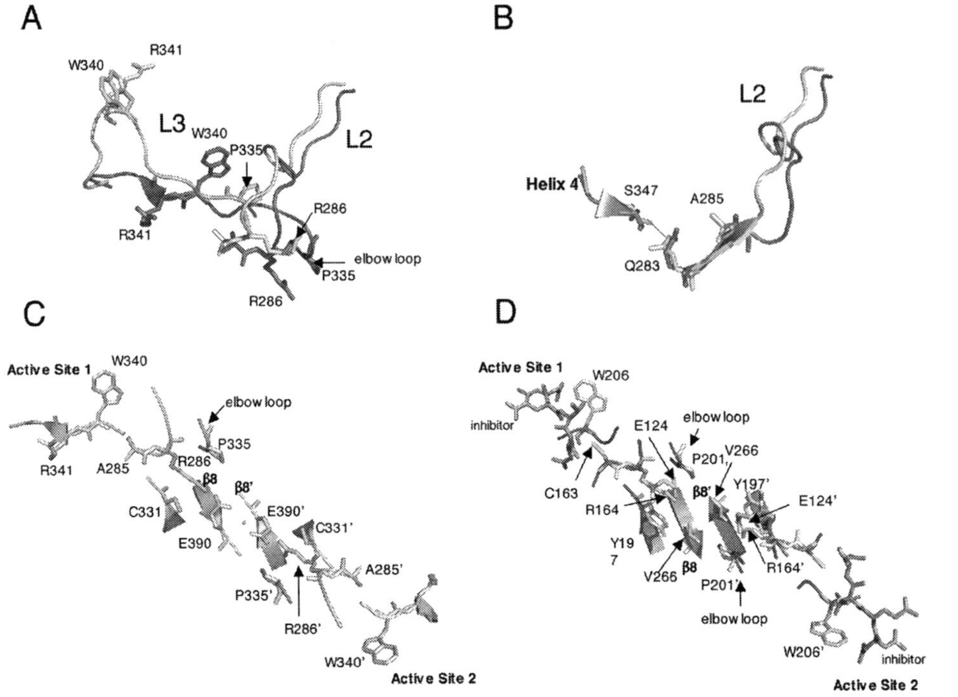

Figure 5. Active site rearrangements of caspases-1 and -3. A) Movements in L2 and L3 of caspase-1 upon substrate binding. The following colors are used. Green: apo-caspase-1 (PDB entry 1SC1), blue: holo-caspase-1 with malonate bound (PDB entry 1SC3). B) Different rotamer configurations for Gln283 in the ligand-free (blue) or ligand-bound (green) conformations of caspase-1 and loop movements in L2. Ala285 refers to the catalytically inactive mutant of caspase-1. C) Movements of Arg286 upon substrate binding result in intercalation of the side chain between Cys331 and Pro335, from active site loop 3, forming a new salt bridge with Glu390 from the dimer interface. The red sphere indicates a water molecule between Glu390 and Glu390'. D) In caspase-3 (PDB entry 2J30), the charge of Arg164 (equivalent to Arg286 in caspase-1) is neutralized by Glu124, which is situated above the dimer interface. For B and C, the secondary structure and active site loops are colored the same as those in Figures 3A and 4A and side chains are colored using the cpk color mode. Structures were generated using *Pymol* (Delano Scientific LLC, Palo Alto, CA). A color version of this image is available at www.landesbioscience.com/curie.

Arg286 is surface-exposed in the ligand-free form but is completely internalized upon ligand binding (Fig. 5A) and is stabilized by formation of the salt bridge with Glu390 (Fig. 5C).

As described above caspase-9 is unique from all other caspases because only one active site has the correct conformation to bind substrate. The other active site is unable to form due to steric constraints of residues in the dimer interface, largely from interactions involving Phe404, Tyr345 and Phe404'/Tyr345' (Fig. 6A and 6B).[32] In order for a functional active site to form, the side chains of Ser344-Ser353, the so-called "elbow loop", from one heterodimer, contact residues in the dimer interface and insert into the opposing monomer (Fig. 6C) and allow the S2 and S1 binding pockets to assume an active conformation. The movement of the active site loops of caspase-9 upon ligand-binding is similar to those of the loops in caspase-1.[48]

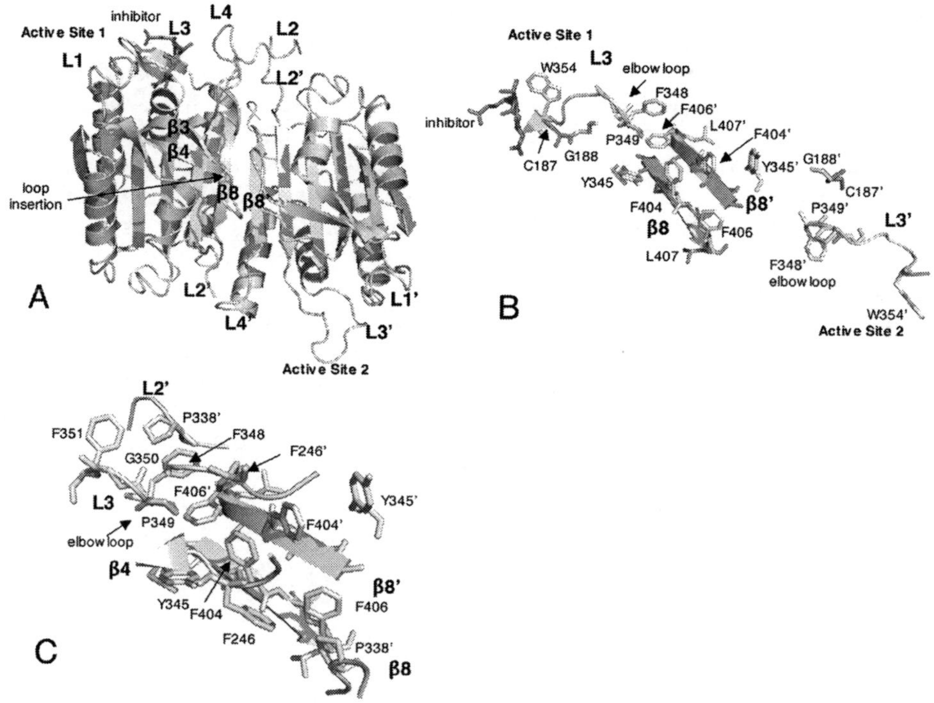

Figure 6. Active site rearrangements of caspase-9 upon inhibitor binding. A) Structure of caspase-9 (PDB entry 1JXQ) with inhibitor bound to one active site. The structure is colored as in Figure 3 except the loop insertion between β-strands 3 and 4 is shown in green. B) Upon inhibitor binding to one active site, loop rearrangements result in movement of the Tyr345 side chain, on β-strand 7, away from the active site and toward the protein interior, causing the side chain of Phe404 to move toward the dimer interface. Due to steric constraints in the interface, these movements can occur only in one heterodimer, so the second active site remains disorganized. C) Interactions among amino acids in the elbow loop (Phe348-Phe351) and the second heterodimer (Phe246', Pro338' and Phe406'). For A-C, the secondary structure and active site loops are colored the same as those in Figures 3A and 4A and side chains are colored using the cpk color mode. Structures were generated using *Pymol* (Delano Scientific LLC, Palo Alto, CA). A color version of this figure is available at www.landesbioscience.com/curie.

ACTIVE SITE COOPERATIVITY

As described above, the interactions at the dimer interface lend to the overall stability of mature caspases. The integrity of the dimer interface also could be responsible for cooperativity between the two active sites. Given that interactions at the dimer interface affect the conformations of the active site loops, the question remained as to whether the binding of substrate to one active site enhanced the binding of substrate to the other active site. This was examined using wild-type caspase-1 and a mutant containing a Glu390 to Ala substitution.[43] The rationale was that Glu390 had a dual role in dimer stabilization and communication across the dimer interface. The results showed that wild-type caspase-1 exhibits limited positive cooperativity with a Hill coefficient of 1.5 and replacing Glu390 with alanine removes the cooperativity (Hill coefficient of 1.0). Current models suggest that Glu390 and

Glu390' interact through a central water molecule in the dimer interface (Fig. 5C) and play a key role in communication between the two active sites. At present, very little is known about the mechanisms of cooperativity in caspase-1 or whether it occurs in other caspases.

In contrast to communications across the dimer interface, the interactions at the dimer interface of one heterodimer are tightly coupled to the productive binding of the substrate to the active site of the same heterodimer. To further illustrate this point, tethering experiments were performed to find allosteric sites for inhibitor binding.[43,49] Tethering uses a library of small thiol-containing compounds that make disulfide bonds with naturally occurring cysteine residues in a protein. This method was used to determine allosteric sites in caspases-3, -7 and -1. Two classes of compounds were identified after screening thousands of molecules. Representatives of these classes are FICA (5-fluoro-1H-indole-2-carboxylic acid (2-mercapto-ethyl) amide) and DICA (2-(2,4-dichlorophenoxy-N-(2-mercapto-ethyl)-acetamide). Both compounds bound to a single cysteine in the small subunit on β-strand 8, Cys264 in caspase-3, which sits close to the dimer interface and is 14 Å from the catalytic cysteine. Binding of the inhibitors did not cause dissociation of the dimer but did prevent binding of the substrate at the active site. Structural data showed that in caspase-7, FICA interacts with Tyr223' (β-strand 7) on the opposing heterodimer to occupy the central cavity

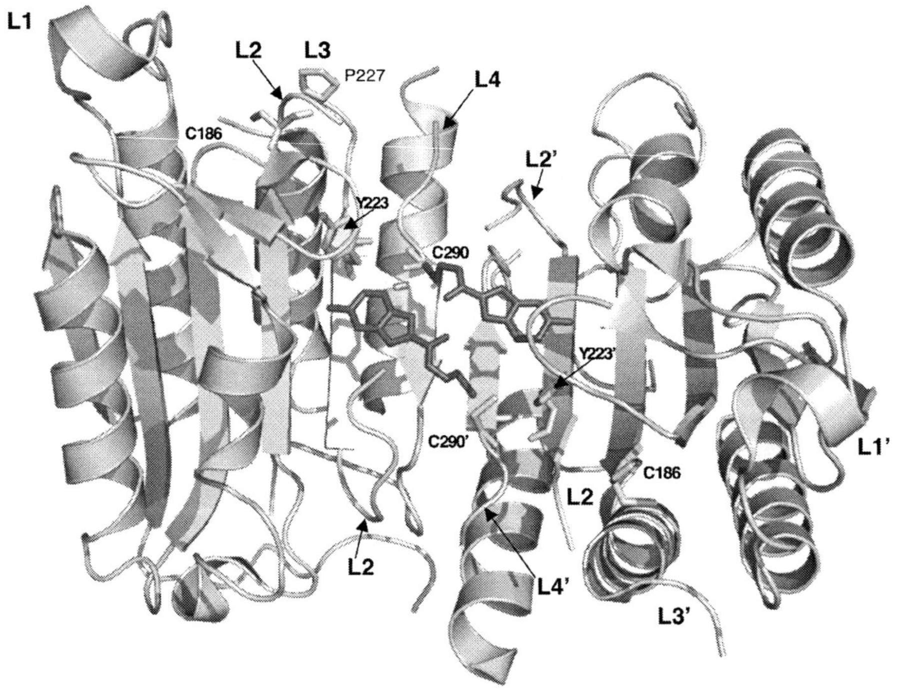

Figure 7. Structure of caspase-7 with FICA bound in the dimer interface (PDB entry 1SHL). Binding of inhibitor in the dimer interface displaces the side chain of Tyr223 and results in disorganized active site loops L2, L3 and L4. L2' occupies the central cavity, as observed for apo-caspase-7 (see Fig. 4B). The secondary structure and active site loops are colored the same as those in Figures 3A and 4A and side chains are colored using the cpk color mode. Structures were generated using *Pymol* (Delano Scientific LLC, Palo Alto, CA).

(Fig. 7), whereas DICA interacts with Tyr223 in the same heterodimer. Regardless, the same conformational changes occur at the active site in the presence of either FICA or DICA. By displacing Tyr223, Arg187 is pushed out of the central cavity and placed into a position that occludes substrate binding to the active site and Cys186 is moved by 3.7 Å away from the S1 subsite. In addition, L2' remains bound over the allosteric site and away from the substrate binding site (Fig. 7). This precludes formation of the loop bundle. As a result, L3 adopts a disordered structure and becomes completely destabilized. In fact, the large conformational changes that occur as a result of inhibitor binding to the interface resemble the conformation of the zymogen rather than that of the active enzyme.

In the case of caspase-1, it was found that a thienopyrazole bound to Cys331, which is near the dimer interface on β-strand 7 (Fig. 5C). Like FICA in caspase-7, the caspase-1 inhibitor interacted with residues on the neighboring heterodimer. Overall, the binding of the tethered inhibitor resulted in a disorganized active site by causing the catalytic cysteine to be rotated 5 Å away from the S1 subsite, the substrate-binding loop to be collapsed so that it could not interact productively with substrate and the side chain of Arg286 to be rotated away from the dimer interface and toward the substrate-binding cleft. As with caspase-7, caspase-1 adopted a conformation closely resembling its ligand-free form when bound to the allosteric inhibitor. Current models suggest that binding of the allosteric inhibitor to the dimer interface shifts the equilibrium of species found in solution from the active form to the inactive form,[43] underlining the coupling of the dimer interface to the active site.

CONCLUSION

In caspases, the composition of the dimer interface is a key determinant of the mode of activation taken by an individual enzyme, in that interfaces that are mostly hydrophobic, like caspase-3, are found as stable, yet inactive, dimers prior to activation and only require processing to achieve a productively active conformation. Dimer interfaces that are more hydrophilic, like those of caspases-1 and -8, tend to result in monomeric caspases in physiological conditions. In those cases, activation requires an increased local concentration to promote dimerization. Overall, dimerization of caspases is a critical event in apoptotic initiation because it ultimately leads to rearrangement of the active site loops, resulting in the formation of the active site and it increases the stability of the caspase. These features ultimately lead to cell death.

NOTE ADDED AFTER PROOF

Recent developments using small molecules to activate procaspase-3 either in allosteric sites or by other mechanisms show promise as a way to therapeutically target apoptosis.[50-53]

ACKNOWLEDGEMENT

This work was supported by a grant from the National Institutes of Health (GM065970).

REFERENCES

1. Jacobson MD, Weill M, Raff MC. Programmed cell death in animal development. Cell 1997; 88:347-354.
2. Fadeel B, Orrenius S. Apoptosis: a basic biological phenomenon with wide-ranging implications in human disease. J Internal Med 2005; 258:479-517.
3. Kabore AF, Johnston JB, Gibson SB. Changes in the apoptotic and survival signaling in cancer cells and their potential therapeutic implications. Current Cancer Drug Targets 2004; 4:147-163.
4. Fulda S, Debatin K-M. Targeting apoptosis pathways in cancer therapy. Current Cancer Drug Targets 2004; 4:569-576.
5. Meng XW, Lee S-H, Kaufmann SH. Apoptosis in the treatment of cancer: a promise kept? Curr Op Cell Biol 2006; 18:668-676.
6. Earnshaw WC, Martins LM, Kaufmann SH. Mammalian caspases: structure, activation, substrates and functions during apoptosis. Ann Rev Biochem 1999; 68:383-424.
7. Wyllie AH, Kerr JF, Currie AR. Cell death: the significance of apoptosis. Intern Rev Cytol 1980; 68:251-306.
8. Kaufmann SH. Induction of endonucleolytic DNA cleavage in human acute myelogenous leukemia cells by etoposide, camptothecin and other cytotoxic anticancer drugs: a cautionary note. Cancer Res 1989; 49:5870-5878.
9. Canman CE, Tange H-Y, Normolle DP et al. Variations in patterns of DNA damage induced in human colorectal tumor cells by 5-fluorodeoxyuridine: implications for mechanisms of resistance and cytotoxicity. Proc Natl Acad Sci 1992; 89:10474-10478.
10. Jin Z, El-Deiry W. Overview of cell death signaling pathways. Cancer Biology and Therapy 2005; 4:139-163.
11. Enari M, Sakahira H, Yokoyama H et al. A caspase-activated DNase that degrades DNA during apoptosis and its inhibitor ICAD. Nature. 1998; 391:43-50.
12. Fuentes-Prior P, Salvesen GS. The protein structures that shape caspase activity, specificity activation and inhibition. Biochem J 2004; 384:201-232.
13. Thornberry NA, Bull HG, Calaycay JR et al. A novel heterodimeric cysteine protease is required for interleukin-1β processing in monocytes. Nature 1992; 356:768-774.
14. Martinon F, Tschopp J. Inflammatory caspases: linking an intracellular innate immune system to autoinflammatory diseases. Cell 2004; 117:561-574.
15. Ahn E-Y, Pan G, Vickers SM et al. IFN-γ upregulates apoptosis-related molecules and enhances FAS-mediated apoptosis in human cholangiocarcinoma. Intern J Cancer 2002; 100:445-451.
16. Boatright KM, Salvesen GS. Mechanisms of caspase activation. Curr Op Cell Biol 2003; 15:725-731.
17. Boatright KM, Renatus M, Scott FL et al. A unified model for apical caspase activation. Mol Cell 2003; 11:529-541.
18. Leung BP, Culshaw S, Gracie JA et al. A role for IL-18 in neutrophil activation. J Immunol 2001; 167:2879-2886.
19. Martinon F, Burns K, Tschopp J. The inflammasome: a molecular platform triggering activation of inflammatory caspases and processing of proIL-β. Mol Cell 2002; 10:417-426.
20. Thornberry NA, Rano TA, Peterson EP et al. A combinatorial approach defines specificities of members of the caspase family and granzyme B. J Biol Chem 1997; 272:17907-17911.
21. Schweizer A, Briand C, Grutter MG. Crystal structure of caspase-2, apical initiator of the intrinsic apoptotic pathway. J Biol Chem 2003; 278:42441-42447.
22. Hofmann K, Bucher P, Tschopp J. The CARD domain: a new apoptotic signalling motif. Trends Biochem Sci 1997; 22:155-156.
23. Thome M, Hofmann K, Burns K et al. Identification of CARDIAK, a RIP-like kinase that associates with caspase-1. Curr Biol 1998; 8:885-888.
24. Weber CH, Vincenz C. The death domain superfamily: a tale of two interfaces? Trends Biochem Sci 2001; 26:475-481.
25. Feeney B, Clark AC. Reassembly of active caspase-3 is facilitated by the propeptide. J Biol Chem 2005; 280:39772-39785.
26. Denault J-B, Salvesen GS. Human caspase-7 activity and regulation by its N-terminal peptide. J Biol Chem 2003; 278:34042-34050.
27. Meergans T, Hildebrandt A-K, Horak D et al. The short prodomain influences caspase-3 activation in HeLa cells. Biochem J 2000; 349:135-140.
28. Cowling V, Downward J. Caspase-6 is the direct activator of caspase-8 in the cytochrome c-induced apoptosis pathway: absolute requirement for removal of caspase-6 prodomain. Cell Death and Differentiation 2002; 9:1046-1056.
29. Muzio M, Stockwell BR, Stennicke H et al. An induced proximity model for capase-8 activation. J Biol Chem 1998; 273:2926-2930.
30. Pop C, Timmer J, Sperandio S et al. The apoptosome activates caspase-9 by dimerization. Mol Cell 2006; 22:269-275.

31. Boatright KM, Renatus M, Scott FL et al. A unified model for apical caspase activation. Mol Cell 2003; 11:529-541.
32. Renatus M, Stennicke HR, Scott FL et al. Dimer formation drives the activation of the cell death protease caspase 9. Proc Natl Acad Sci 2001; 98:14250-14255.
33. Li H, Bergeron L, Cryns V et al. Activation of caspase-2 in apoptosis. J Biol Chem 1997; 272:21010-21017.
34. Baliga BC, Read SH, Kumar S. The biochemical mechanism of caspase-2 activation. Cell Death and Differentiation. 2004; 11:1234-1241.
35. Launay S, Hermine O, Fontenay M et al. Vital functions for lethal caspases. Oncogene 2005; 24:5137-5148.
36. Bose K, Clark AC. Dimeric procaspase-3 unfolds via a four-state equilibrium process. Biochemistry 2001; 40:14236-14242.
37. Bose K, Clark AC. pH effects on the stability and dimerization of procaspase-3. Protein Sci 2005; 14:24-36.
38. Matsuyama S, Llopis J, Deveraux QL et al. Changes in intramitochondrial and cytosolic pH: early events that modulate caspase activation during apoptosis. Nat Cell Biol 2000; 2:318-325.
39. Bose K, Pop C, Feeney B et al. An uncleavable procaspase-3 mutant has a lower catalytic efficiency but an active site similar to that of mature caspase-3. Biochemistry 2003; 42:12298-12310.
40. Mittl PRE, DiMarco S, Krebs JF et al. Structure of recombinant human CPP32 in complex with the tetrapeptide acetyl-asp-val-ala-asp fluoromethyl ketone. J Biol Chem 1997; 272:6539-6547.
41. Ganesan R, Mittl PRE, Jelakovic S et al. Extended substrate recognition in caspase-3 revealed by high resolution X-ray structure analysis. J Mol Biol 2006; 359:1378-1388.
42. Richardson JS, Richardson DC. Natural β—sheet proteins use negative design to avoid edge-to-edge aggregation. Proc Natl Acad Sci 2002; 99:2754-2759.
43. Scheer JM, Romanowski MJ, Wells JA. A common allosteric site and mechanism in caspases. Proc Natl Acad Sci 2006; 103:7595-7600.
44. Blanchard H, Kodandapani L, Mittl PRE et al. The three-dimensional structure of caspase-8: an initiator enzyme in apoptosis. Structure 1999; 7:1125-1133.
45. Chai J, Wu Q, Shiozaki E et al. Crystal structure of a procaspase-7 zymogen: Mechanisms of activation and substrate binding. Cell 2001; 107:399-407.
46. Riedl SJ, Fuentes-Prior P, Renatus M et al. Structural basis for the activation of human procaspase-7. Proc Natl Acad Sci 2001; 98:14790-14795.
47. Wilson KP, Black J-AF, Thomson JA et al. Structure and mechanism of interleukin-1β converting enzyme. Nature 1994; 370:270-275.
48. Romanowski MJ, Scheer JM, O'Brien T et al. Crystal structures of ligand-free and malonate-bound human caspase-1: implications for the mechanism of substrate binding. Structure 2004; 12:1361-1371.
49. Hardy JA, Lam J, Nguyen JT et al. Discovery of an allosteric site in the caspases. Proc Natl Acad Sci 2004; 101:12461-12466.
50. Walters J, Pop C, Scott FL et al. A constitutively active and uninhibitable caspase-3 zymogen efficiently induces apoptosis. Biophys J 2009; 424:335-345.
51. Schipper JL, MacKenzie SH, Sharma A, Clark AC. A bifunctional allosteric site in the dimer interface of procaspase-3. Biophys Chem 2011; 159:100-109.
52. Wolan DW, Zorn JA, Gray DC, Wells JA. Small molecule activators of a proenzyme. Science 2009; 326:853-858
53. Peterson QP, Goode DR, West DC, Ramsey KN, Lee JJ, Hergenrother PJ. PAC-1 activates procaspase-3 in vitro through relief of zinc-mediated inhibition. J Mol Biol 2009; 388:144-158.

CHAPTER 5

THE RELATIONSHIP BETWEEN OLIGOMERIC STATE AND PROTEIN FUNCTION

Michael D.W. Griffin[1] and Juliet A. Gerrard*[,2]

[1]Bio21 Institute of Molecular Science and Biotechnology, University of Melbourne, Melbourne, Victoria, Australia; [2]Biomolecular Interaction Centre (BIC), School of Biological Sciences, University of Canterbury, Christchurch, New Zealand

**Corresponding Author: Juliet A. Gerrard—Email: juliet.gerrard@canterbury.ac.nz*

Abstract: The reason that many proteins adopt a particular oligomeric form is far from obvious. In this chapter, we discuss potential advantages of proteins self-assembling into specific quaternary structures. A number of case studies are presented in which wild-type proteins have been mutated to generate variants of lower oligomeric order and the impact on the resulting proteins, in terms of both specific function and generic stability, are discussed. Drawing on these case studies, some general design principles for quaternary structure engineering are put forward to facilitate these experiments on a wider range of systems. It is clear that the advantages afforded by quaternary structure vary from protein to protein; however, some general trends are starting to emerge.

INTRODUCTION

While the existence of multisubunit proteins is well documented, our understanding of the principles behind the supramolecular assembly of proteins into quaternary complexes remains rudimentary compared to our sophisticated knowledge of protein tertiary structure. The question of why such a large proportion of proteins are composed of subunits in this way has stimulated great interest in the delineation of the structural and functional roles of subunits in oligomeric proteins;[1] however, knowledge in this field has progressed little since early work in the area.[2] Understanding the functional roles of protein oligomerisation has proved difficult, in part because of the relative paucity of biophysical data: the proportion of known proteins that are well characterised in terms of their quaternary structure is notably low.[3]

Protein Dimerization and Oligomerization in Biology, edited by Jacqueline M. Matthews.

Investigations of the biophysics and mechanistic contributions of intersubunit interactions are, therefore, critical to our understanding of how the quaternary structures of these proteins endow them with their unique biological properties.[1] One way to probe the role that the oligomeric state plays in a protein's function is to engineer proteins of different oligomeric order and compare their structure and function to the wild-type protein. Studies of this sort have the potential to provide insight, not only into the details of specific multimeric systems, but also into characteristic factors that may be applied to broad classes of proteins by relation of tertiary and quaternary structural features. This chapter therefore highlights the current literature concerning the role of quaternary structure and insights obtained via quaternary structural engineering.

THE QUATERNARY COMPLEX AND PROTEIN-PROTEIN INTERACTIONS

The self-association of proteins to form quaternary complexes (oligomers) is a very common phenomenon. It has been estimated that the average oligomeric state of cellular proteins is tetrameric.[4] 70-80% of proteins are comprised of a number of monomeric subunits that associate noncovalently to form functional oligomeric proteins.[5-7] These multimeric proteins vary enormously in complexity, ranging from dimers to giant holo-enzymes composed of large numbers of subunits[1] and each may consist of identical monomers, or two or more different types of subunit. However, the large majority of multimeric proteins are either homodimeric, consisting of two identical monomers, or homotetrameric, consisting of four identical monomers. A recent poll of the Brenda enzyme database revealed that of 311 characterised oligomeric human proteins, 199 form homo-multimers, with homodimers comprising 63% (125) of these and homotetramers 25% (50).[3] (See also Chapter 1, by Matthews and Sunde.)

Homo- and Hetero-Complexes

Protein quaternary complexes can be composed of identical (homo-) or non-identical (hetero-) protein units. In the case of homo-complexes, these protein units can be classified as being organised in an isologous or heterologous manner,[8,9] with associated symmetry[10,11] (Fig. 1). An isologous association involves the same surface and hence the same contact residues, of both monomers, which are related by a two-fold symmetry axis. Further oligomerisation of an isologous association can only be achieved by a different interface. In contrast, heterologous assemblies involve association of non-identical surfaces, which, without closed, cyclic, symmetry, can lead to infinite aggregation. By their nature, hetero-oligomers must form heterologous associations. This classification, first introduced by Monod,[16] remains useful today; although it may not adequately incorporate certain oligomeric interactions, such as domain swapping.[4]

Obligate and Non-Obligate Associations

Multimeric protein structures are also distinguished on the basis of whether a complex is obligate or non-obligate.[9] An obligate interaction is usually very stable and thus, normally only exists in its complexed form; hence, the components of the complex are not found as isolated stable structures in vivo.[9] These complexes are generally also thought to be

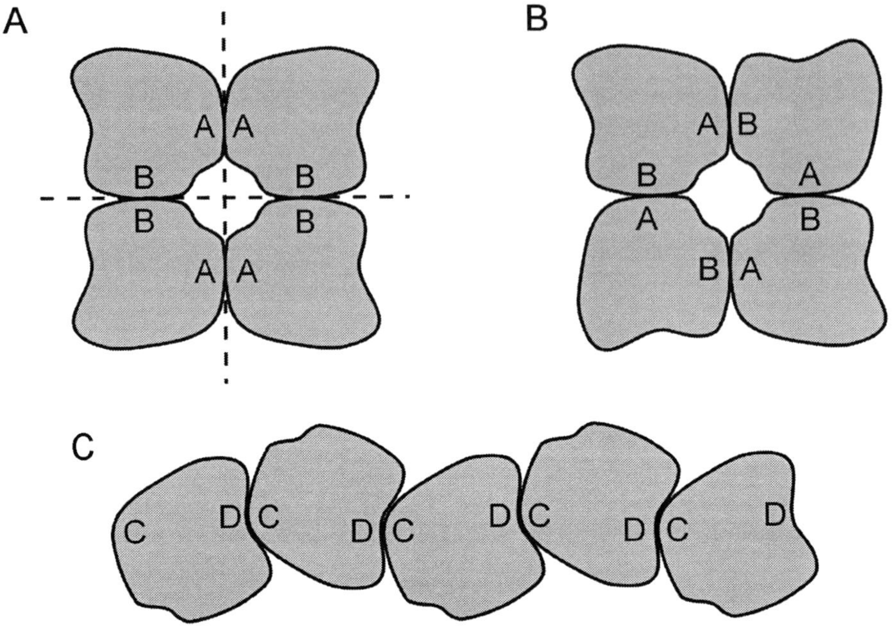

Figure 1. Examples of isologous and heterologous arrangements of identical subunits. A) Homotetramer with isologous interfaces. Two two-fold symmetry axes are shown as dashed lines, the third is orthogonal to the page at the centre of the structure. B) Homotetramer with heterologous interfaces. C) Heterologous association without cyclic symmetry allows unrestricted aggregation.

functionally obligate; however, this may not always be the case. Non-obligate multimeric structures form complexes that associate and dissociate in vivo. These complexes can be further classified as those in which weak interactions produce a dynamic oligomeric equilibrium in solution and those in which stronger transient interactions require a molecular or physiological trigger to alter the oligomeric state.[9] Of course, these classifications are not mutually exclusive, as equilibrium between varying oligomeric orders is a feature of all multimeric structures; however, the position and nature of the equilibrium may play an important role in protein function. (See also Chapter 3, by Jones.)

Structural Characteristics of Protein-Protein Interfaces

The increasing volume of structural data available for various protein quaternary complexes has allowed comparative analysis of the structural characteristics of protein-protein interfaces. Most recently, the interfaces of wide ranges of protein oligomers have been assessed by parameters such as the size and shape of the contact area, protrusion or flatness of the interface, the chemical and physical properties of the surface and the amino acid composition of contact residues[12-19] (see also Chapter 3, by Jones). The residues at an oligomeric interface may be more conserved than other surface residues, especially so-called protein 'hot-spots' that often rest at the centre of the interface and account for a large proportion of the binding energy for the interaction.[4] This has led to limited success in prediction of protein interface sites from structural and sequence data.[14,20-22] From these studies, it is possible to extract a number of general features of protein-protein interfaces.

The interfaces of both hetero-complexes and homo-complexes are generally circular and planar,[3,15,16] and a clear, although scattered, correlation exists between the size of the interface and the molecular weight of the monomer, with larger subunits generally having larger interfaces.[15,16] However, the overall binding energy between monomers does not appear to correlate with the size of the interface.[19,23] The interfaces of obligate complexes are generally larger and more hydrophobic in nature than those involved in non-obligate complexes,[15,16,18] which generally exhibit a more polar interface, presumably to meet the requirements for independent monomer folding and solubility. Oligomeric structures with interface areas of less than 1,000 Å^2 are unlikely to undergo large scale conformational changes on complex formation.[9,18,19]

Evolution of Protein-Protein Interfaces

The structure and affinity of a protein-protein association is tuned to its biological function and/or mechanism.[9] Thus, protein-protein interfaces presumably evolve to optimise association with respect to functional efficacy. Although it has not been clearly established how specific protein-protein associations have evolved, it is reasonable to assume that weak, nonspecific interactions that confer a functional advantage for an organism may be expanded by fine tuning of the physics of association and thus evolve into specific associative interfaces.[3] This evolutionary process, as any, requires significant selective pressure, implying that the advantage conferred must outweigh the disadvantages of the evolutionary process. Thus quaternary complexes are unlikely to evolve 'by accident'.

THE ROLES OF QUATERNARY STRUCTURE IN PROTEIN FUNCTION

A range of factors has been proposed to account for the high incidence of multimeric proteins. The most familiar of these are the facilitation of regulatory or substrate induced cooperativity between the subunits of an oligomer and the formation of shared active sites[1,3] (Fig. 2). Structural proteins that form large filamentous or scaffold structures (Fig. 2A) within the cell present a special case of a functional consequence of oligomerisation that is beyond the scope of this discussion.

Specific Functional Advantages

The haemoglobin tetramer provides the archetypal example of a protein that undergoes structural change on ligand binding to any monomer, leading to the generation of a tetrameric conformation with increased ligand binding affinity[24] (Fig. 2B). The delineation of these complex mechanisms of concerted cooperativity represents the original demonstration of the requirement of specific quaternary structure for biological function and a range of multimeric proteins have since been found to display this sort of regulatory mechanism. However, these proteins are predominantly hetero-complexes, composed of non-identical monomers and account for only about 30% of all oligomeric proteins.[6,7]

Shared active sites, in which the active site is formed at the interface of two monomers of an oligomeric enzyme and is contributed to by both monomers have frequently been observed[3] (Fig. 2C). This feature places enzyme activity under the regulation of

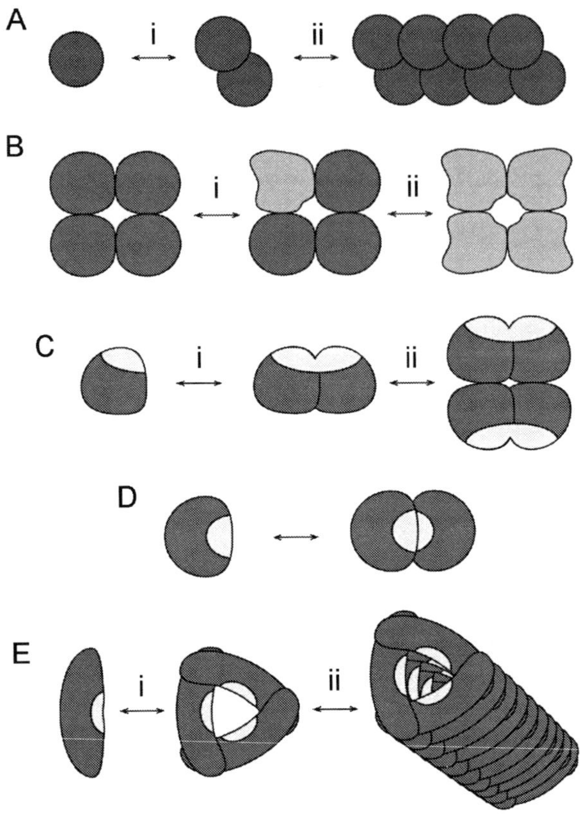

Figure 2. Functional consequences of protein oligomerisation. A) Dimerisation of the monomer (i), allows aggregation of stable structural elements in the cell (ii). B) Binding of a cofactor or substrate to a single subunit of an oligomer can alter the conformation of that subunit (i) and can cooperatively induce structural changes in the remaining subunits (ii). C) Oligomerisation can generate complete active or binding sites at the interface (i). Further association of these dimers may have no direct role in active site construction (ii). D) Alternatively, oligomerisation can block existing active or binding sites of the monomer or E) create a supramolecular structure that increases enzyme activity by channelling of substrate. Adapted from Marianayagam et al. Trends Biochem Sci 2004; 29(11):618-625.[3]

oligomerisation: if the protein concentration drops below the threshold of association, or some molecular trigger causes disruption of the oligomeric structure, inactive monomers are formed.[3] Similarly, monomer-monomer association may cause blocking of the active site. In this case, the monomer is the active species (Fig. 2D).

An elegant example of active site sharing can be found in the dimeric Type II restriction endonucleases, which bind and cleave palindromic DNA sequences. The binding sites of each monomer of the homodimer are arranged symmetrically across the interface such that each subunit binds half of the palindrome.[3] Whereas active site sharing can provide a rationale for monomer-monomer association, it may not explain the full order of the quaternary complex, for example, in the case of *Escherichia coli* dihydrodipicolinate synthase, a tight dimer structure is required for full configuration of

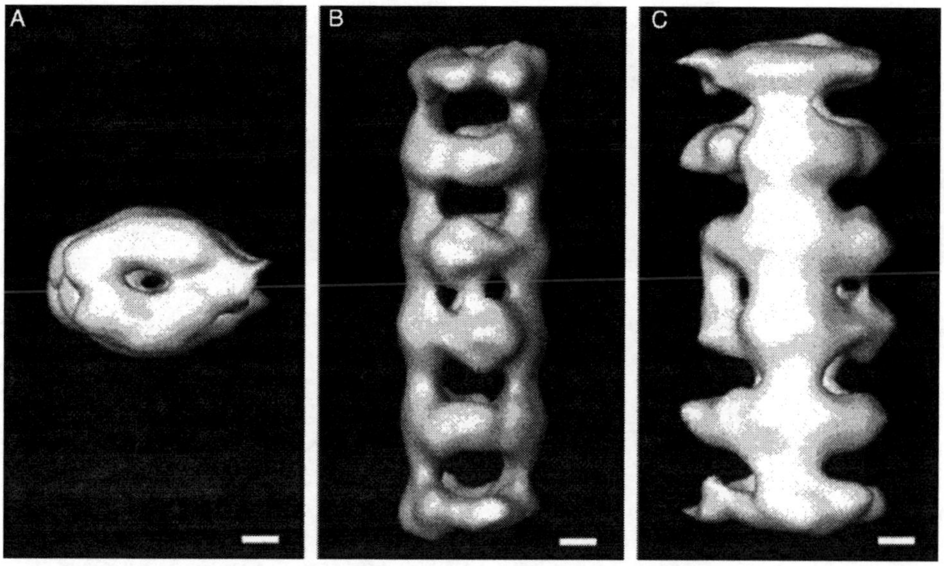

Figure 3. Three-dimensional reconstruction of the oat β-glucosidase multimers derived from a tomographic tilt series. A), top view. B,C) different side views. Scale bar corresponds to 2 nm. Reproduced with permission from Kim SY et al. J Struct Biol 2005; 150(1):1-10.[26]

active site and regulatory motifs, but the association of two dimers to form the tetramer plays no direct role in active site construction (Fig. 2C).[25]

Multimerisation may also play a role in channel formation: oat β-glucosidase has been shown to assemble into a long fibrillar assembly with a unique quaternary structure comprised of stacked trimeric units (Fig. 3).[26] The active sites of the enzyme are located inside the tunnel (Fig. 2E), increasing the affinity for the substrates and potentially enhancing specificity of action.[26] As more cryo-electron micsroscopy studies of this nature are carried out, it will be interesting to see whether such supramolecular protein assemblies and by implication substrate channelling, prove to be a general feature in vivo.

Generic Advantages

In other oligomeric proteins, most notably those composed of identical subunits, the functional properties conferred by the quaternary structure are not immediately obvious from structural data or functional studies.[1] In these cases, a range of generic benefits have been ascribed in an effort to justify their existence as quaternary complexes. These include factors relating to the increased size of the complexes relative to the monomers and economy of scale with respect to protein synthesis and genome size.[10] Burial of hydrophobic domains within interfaces that would otherwise be exposed, leading to insolubility, has been put forward as a possible justification for the formation of oligomeric structures.[1] Consideration of the evidence presented earlier in this chapter would suggest that the hydrophobic nature of the contact surfaces of obligatory quaternary complexes is a consequence of, rather than a cause of, the formation of these interfaces.[4] In order for a protein to develop higher order oligomers, it must itself have an intrinsic level of stability and solubility.

Large proteins with extensive internal interactions generally have more stable folded structures than small proteins.[10] This is because in large proteins the enthalpic contribution of the many weak internal interactions far outweighs the entropic cost of the constrained conformation, whereas in smaller proteins this balance is not as pronounced.[10] The oligomerisation of multiple identical monomers provides a relatively simple and economical method of forming large protein structures and may be a general means for increasing stability against denaturation. Here it must be noted that the effect of oligomerisation on the stability of a protein is highly dependent on the characteristics of the interface and the nature of the interactions between monomers. A number of cases have been documented in which monomeric forms of obligate quaternary complexes show equal or increased stability when compared to the oligomeric structure.[27-30]

Another facet of stability that may be offered by oligomeric structure is protection against misfolding and formation of potentially pathogenic protein aggregates such as amyloid fibrils, which have been implicated in a number of neurodegenerative diseases.[31] In the amyloidogenic proteins insulin, κ-casein, immunoglobulin light chain and transthyretin, it has been shown that break down of the natural oligomeric complexes into monomers is necessary for nucleation of amyloid fibrils,[32-36] and that noncovalent stabilisation of the oligomeric structures inhibits amyloid formation.[37] Further, hereditary mutations in the tetrameric human transthyretin that cause disruption of the tetramer into dimers[38] or monomers[39] increases the propensity of the protein to form amyloid, even where the stability of the monomeric units is unaltered.[38] Thus one could argue that oligomeric proteins are 'self-chaperoned', with each monomer protected from the risks of unfolding and 'nonnative' associations by 'safe' native interaction with other monomers.

The increased size of oligomeric complexes with respect to the free constituent monomers also has implications for the water regulation of a cell or organelle. Larger protein structures have lower membrane permeability and therefore reduced osmotic pressure.[1] Formation of complexes reduces the available protein surface area, thereby significantly reducing the amount of water required for full hydration of the protein.[10] These factors may not present a major evolutionary driving force, however, as it has been shown that many cell types can loose over half of their water without adverse effects.[10,40]

Related to these considerations of size is the error management and efficiency of coding offered by homo-oligomeric proteins.[1,10] As the protein is made up of a number of polypeptide chains, translation errors may be reduced by discarding individual protein chains with sequence defects, as opposed to rejecting one large chain. Also, the assembly of a number of identical protein monomers into a quaternary complex allows creation of a large structure using the minimum amount of genetic material. However, the large amount of noncoding DNA in eukaryotic genomes suggests that this may not be a driving force in higher organisms.

A further generic advantage of a particular oligomeric state may arise from the need to optimise the dynamics at the active site of a monomer. This has recently been postulated for *E. coli* dihydrodipicolinate synthase (see Griffin et al., 2008)[88] but remains to be tested in other systems.

QUATERNARY STRUCTURAL ENGINEERING

Certain attributes of oligomeric proteins, for example determining requirements to encode intersubunit interfaces, can be probed using mini-protein models, as recently reviewed.[4] Insights into the role of quaternary structure in natural proteins have also been obtained via structural engineering of the interfaces of oligomeric proteins, considered here.

Engineering of Thermostable Interfaces

Interest in the mechanisms by which proteins from thermophilic bacteria achieve their remarkable thermal tolerance prompted examination of the characteristics of the intersubunit interfaces of the proteins form these organisms.[41,42] It was found that increased density of hydrophobic packing and more extensive ion-pair networks were commonplace in the intersubunit interfaces of these proteins when compared to their nonthermally resistant homologues.[43,44]

Applying these principles, the thermal stabilities of hexameric glutamate dehydrogenase from two bacterial species[45,46] and the non-obligate dimeric TATA-binding protein from yeast[47] were significantly enhanced by the introduction of amino acid residues forming ion-pair networks within the interfaces of these proteins. Improvement of the hydrophobic packing within the interface of the dimeric Lac repressor of *E. coli* has resulted in a remarkable increase in thermostability from 47°C to 87°C.[44] Similarly, mutation of residues at the dimer interface of yeast triosephosphate isomerase based on their chemical properties and contribution to association of the two monomers doubled the half life of the enzyme at 100°C.[48,49] These studies suggest that oligomeric structure may play a role in thermal stability.

Engineering of Lower Order Quaternary Structures from Existing Oligomers

The literature contains a limited number of reports of studies directed towards altering the quaternary structure of oligomeric proteins by site-directed mutagenesis. These investigations have been undertaken for a number of reasons, including examination of the functional properties of proteins and production of improved clinical therapeutic agents.

Early Studies—Tyrosyl t-RNA Synthetase

The earliest reported investigation into the possibility of dissociation of an obligate oligomeric protein into stable, independent monomers by site-directed mutagenesis of interface residues was carried out by Jones et al.[50] The study focussed on the homodimeric enzyme tyrosyl t-RNA synthetase from *Bacillus stearothermophilus*, which contains two complete and separate active sites, but binds only one equivalent of tyrosine and one equivalent of t-RNA per dimer. This half-of-sites reactivity was presumed to involve negative cooperativity of substrate binding between the two monomers, but had no known mechanistic function. Protein engineering was therefore applied in an effort to create stable and active monomeric units of the enzyme by disruption of the dimeric structure and thereby shed light on the functional contribution of the protein complex.

Disruption of the monomer-monomer interface was achieved by introduction of negatively charged groups at previously interacting positions in the interface, thus producing electrostatic repulsion between the two subunits at this location. The mutant form of t-RNA synthetase produced by this strategy was found to display pH dependent dissociation into stable monomers, consistent with the ionisation of the carboxyl groups producing repulsion. The monomeric form of the enzyme was found to be inactive and was therefore not amenable to the kinetic characterisation necessary for functional analysis. However, subsequent work, in which active heterodimers consisting of this monomeric mutant complexed with a truncated form of the enzyme with an engineered complementary charge at the interface,[51,52] assisted in the elucidation of the binding mode of t-RNA with the dimer.[53]

Insulin

The implications of quaternary structural engineering for the pharmacokinetics of insulin administration in diabetics were quickly seized upon.[54,55] The treatment of Type I diabetes requires the subcutaneous injection of the hormone insulin at meal times to control resultant blood glucose levels. Human insulin exists largely as a homohexamer, composed of a trimer of dimers and, therefore, exerts a relatively low osmotic pressure. This resulted in slow absorption of insulin from the subcutaneous tissue into the bloodstream and slow clearance of blood insulin levels, resulting in abnormal diurnal insulin patterns.

It was recognised that smaller forms of the hormone were likely to be absorbed more rapidly into the bloodstream. Monomeric forms of human insulin were produced by engineering single and double mutations at the dimer interface of the monomer.[54,55] These monomeric analogues were found to be absorbed up to three times faster than the wild-type insulin and show a more natural clearance profile. It was also found that most of these mutations did not significantly effect the in vivo potency of the hormone.[55] As a result of these studies, a number of engineered monomeric insulin analogues are commercially available and in common use today.[56,57] It is curious that a variant with altered quaternary structure has proved more successful clinically than the wild-type protein, hinting that the evolution of the native hexamer may have been driven by factors other than physiological function. Given the predilection of insulin to form amyloid fibrils in vitro,[37] it is tempting to speculate that protection of the monomer from aggregation may have been a driving force for oligomerisation of insulin.

Serratia Marcessens Endonuclease

The design of monomeric variants of the *Serratia marcessens* nonspecific endonuclease was originally directed towards commercial applications, as the enzyme is utilised for removal of nucleic acids from biochemical and pharmaceutical preparations.[27,58] Efficient immobilisation of the wild-type homodimer proved problematic and thus a stable, active monomer was engineered.[27] Detailed functional analysis of the resulting monomeric forms revealed an interesting mechanistic advantage afforded by the dimeric structure of the enzyme.

The monomeric form of the nuclease showed the same specific activity as the wild-type dimer when polynucleotide substrate concentration was high. However, as substrate concentration became limiting, the dimer showed relatively higher activity.[59] The two active sites of the wild-type *S. marcessens* nuclease dimers are separate and operate independently. Thus, a binding event at one active site will raise the apparent local concentration of substrate for the other binding site of the dimer due to protruding nucleic acid strands either side of the cleavage site. This raises the probability of a binding event at the other active site. Thus, a processive mechanism of nucleic acid degradation is produced when substrate concentration is low,[59] allowing more efficient degradation of nucleic acids.

λ Cro

A monomeric form of the homodimeric DNA binding protein λ Cro, produced by extension of the C-terminal, β-strand forming dimerisation domain, was found to have half operator binding specificity.[28] Interestingly, this mutant, in which two β-turn residues were optimised for β-strand formation, is significantly more stable than the wild-type dimer, but with impaired function. Thus there seems to be no general relationship between oligomeric state, function and stability per se. Work on retroevolution of λ Cro generated monomers that retained the wild-type fold and relatively low dimerisation propensities.[60]

Triosephosphate Isomerase

The homodimeric enzyme triosephosphate isomerase (TIM) has undergone rigorous characterisation by quaternary structural engineering. The report of the crystal structure of recombinant derived trypanosomal TIM[61] also described a study in which an interface histidine residue, forming water-mediated hydrogen bonds between the two monomers, had been mutated producing a dimer that dissociates at low protein concentration and displays concentration dependent activity.[61,62] This study was quickly followed up by the creation of an obligate monomeric form of the enzyme, designated monoTIM, by shortening of the 15 residue major interface loop.[29] This monomer was catalytically active; however, the k_{cat} was approximately 1000-fold lower than the wild-type enzyme and was independent of protein concentration. The crystal structure of monoTIM indicated that this large reduction in activity was likely to be due to the reordering of two minor interface loops which contain two of the four essential catalytic residues.[63] This indicates that the dimeric state of the enzyme is necessary for maintenance of the correct conformation of these loops and therefore the positioning of these essential residues. The requirement for TIM to be dimeric for full activity, coupled with some species specific variation at the interface, has enabled small molecules to be designed that disrupt the interface and have potential as antitrypanosomatid drugs.[64] Indeed, the increasing number of proteins found to be inactive if the quaternary structure is disrupted has led to considerable recent attention in the use of protein-protein interface disruptors as drugs.[65-67]

Subsequent introduction of more conservative double point substitutions of interface residues of trypanosomal TIM resulted in an alternative monomeric variant,[68] with kinetic and physical properties similar to those of monoTIM. Further single and double point mutant monomeric forms were crystallised, in the presence and absence of substrate analogue inhibitors.[69] The resulting structures showed that, on complexation with the inhibitors, the reordered catalytic residue containing loops adopt conformations similar to those observed in the wild-type enzyme, suggesting that it is the rigidity of these loops, resulting from interface formation, that is important for optimal catalysis.[69]

In contrast to these studies, where the thermal stability of monomeric variants of trypanosomal TIM was found to be similar to that of the wild-type,[29,68] an obligate monomeric variant of human TIM produced by double point substitution was shown to have greatly reduced conformational stability.[70] Detailed fluorescence spectroscopic denaturation analysis indicated a conformational stability of 2.5 kcal/mol for the monomer, reduced from 19.3 kcal/mol for the wild-type dimer. However, subsequent introduction of α-helix stabilising mutations produced a partial return of conformational stability of the monomer, to 3.9 kcal/mol.[71]

Table 1. Summary of published quaternary structural engineering studies. This list is not exhaustive but provides an overview of many of the literature reports.

Protein	Natural Oligomeric Form	Engineered Form	Effect
Bacillus stearothermophilus tyrosyl t-RNA synthetase	Homodimer	Monomer	Inactive.[50] Reconstruction of active heterodimers assisted in elucidation of t-RNA binding mode.[51-53]
Human insulin	Homohexamer (trimer of dimers)	Monomer	Retained biological activity. Shows dramatically improved pharmacokinetics after subcutaneous injection.[54,55]
Human interleukin-5	Homodimer (domain-swapped)	Monomer	Lacked biological activity.[72] Extension of loop constraining domain-swapped helix resulted in full unimolecular folding and return of biological activity.[73]
Serratia marcessens endonuclease	Homodimer	Monomer	Retained activity.[27] Elucidated 'processive' mechanism of substrate cleavage.[59]
Streptavidin/avidin	Homotetramer	Dimer[74,75] and monomer.[76]	Retained relatively high affinity for biotin.[76] Mutants tetramerise at high concentration.[75]
E. coli inorganic pyrophosphatase	Homohexamer (dimer of trimers)	Trimer,[77] dimer and monomer.[78]	Progressive loss of catalytic efficiency with decreased oligomer order.[78]
EcoRI endonuclease	Homodimer	Monomer	Nonspecific nickase activity; proposed to be due to inability of monomer to bend DNA.[79]
*Bse*634I endonuclease	Homotetramer	Dimer	Retained wild-type catalytic efficiency. Stability compromised.[80]
λ Cro	Homodimer	Monomer	Retained half operator binding specificity. Significantly more stable than wild-type dimer.[28]
Trypanosomal triosephosphate isomerase	Homodimer	Monomer	Activity of original mutant reduced 1000 fold.[29] More conservative mutations produced monomer with similar activity to wild-type.[68]
Human triosephosphate isomerase	Homodimer	Monomer	Greatly reduced activity and conformational stability.[70]

continued on following page

Table 1. Continued

Protein	Natural Oligomeric Form	Engineered Form	Effect
Human superoxide dismutase	Homodimer	Monomer	Activity reduced 10 fold. Showed increased disorder at interface surfaces.[81,82]
Rat prostatic acid phosphatase	Homodimer	Monomer	Inactive.[83]
E. coli Malate dehydrogenase	Homodimer	Monomer	Activity reduced 1.4 to 17,500 fold in different mutants.[84]
Rabbit fructose-1,6-bisphosphate aldolase	Homotetramer	Dimer[85] and monomer[86]	Retained activity similar to wild-type despite reduced stability.[85,86]
Bacillus stearothermophilus Lactate dehydrogenase	Homotetramer	Dimer	Retained activity and thermal stability similar to wild-type; insensitive to allosteric activator.[30]
E. coli Succinyl CoA synthetase	Heterotetramer ($\alpha\beta$-dimer)	Dimer ($\alpha\beta$)	Retained wild-type activity.[87]

Reduced Activity in Quaternary Structural Engineered Proteins

Reduction of catalytic activity has been a common observation in studies of the alteration of the quaternary structure of enzymes (Table 1). Often this attenuation of activity has been difficult to explain and has, therefore, been ascribed to reduced stability in the protein structure. It is then easy to reach the conclusion that the primary function of the quaternary structure of these enzymes is to confer generalised structural stability and consequentially catalytic efficiency. However, contrary to this, cases have been described where reduced stability does not correlate with reduced activity and vice versa (Table 1). Furthermore, studies in which different mutational strategies (see below) were employed for a single protein resulted in mutants of the same quaternary structure but varying stability and activity. These results illustrate the need for careful design and interpretation of this type of experiment.

RATIONAL DESIGN OF MUTATIONS TO DISRUPT THE INTERFACES OF PROTEINS AND CREATE FORMS OF LOWER OLIGOMERIC ORDER

The disruption of the interfaces of a multimeric protein with the aim of producing stable, folded and active protomer units is not a trivial task. The planning of mutations designed to disrupt the quaternary structure of an enzyme requires the consideration of three well-defined problems.[74]

First, subunit-subunit association must be prevented, preferably by introduction of the least possible alteration to the primary structure of the protein. Thus, the smallest number of mutations that is able to produce the desired result of disallowing the associative interface to form must be used to ensure that possible resultant secondary or tertiary structural changes are kept to a minimum. Related to this consideration is the nature of the amino acid substitutions utilised; clearly, it is possible to minimise any structural disturbance to subunits by careful selection of structurally similar amino acids for substitutions. Isosteric replacements are more desirable than otherwise in this respect.

Second, substitutions must be selected that will have minimum impact on any ligand binding or active sites of the protein. Potential positions for alteration must be selected that are distal to any effector binding site or active site to minimise the possibility that potential structural rearrangements caused by the mutation are transmitted to these sites.

Third, separation of the subunits of an oligomeric protein exposes to solvent the side chains of a number of hydrophobic amino acid residues that would normally be buried within subunit-subunit interfaces. This has implications for the aqueous solubility of the subunits and sufficient solubility of the mutant may only be achieved by significant reduction of the hydrophobicity of the exposed interfaces.

Among the published studies described above four different approaches have been taken to destabilising subunit interfaces.

Loop Deletion/Insertion

In many oligomeric proteins, amino acid residues that form the intermonomer interfaces are located on extended loop regions. By replacing these regions with alternative shorter fragments, associative interactions between subunits can be eliminated. Alternatively, by lengthening the loop region, association of two monomers can be blocked by steric hindrance.

Small Residues Mutated to Large Residues

By mutating small amino acid residues that form tightly packed contacts in the subunit interfaces to more bulky amino acids, for example, tyrosine, the topology of the interface surface can be altered, reducing complementarity and disallowing the formation of attractive interactions between subunits.

Hydrophobic Residues Mutated to Hydrophilic Residues

In cases where subunit interfaces are composed mainly of hydrophobic contacts, simply mutating these residues to hydrophilic amino acids can destabilise the interface sufficiently to cause dissociation.

Charged Residue Substitution

This approach uses the property of electrostatic repulsion of like charge. Interface contact residues are mutated to amino acids with ionisable side chain groups that are placed in positions adjacent to residues of the same charge in the neighbouring subunit (if the interface were allowed to form) thus creating electrostatic repulsion between the two subunits.

CONCLUSION

Our understanding of the quaternary structure of proteins remains in its infancy, compared to our much more comprehensive knowledge of how individual proteins fold into their tertiary structures. However, a few general principles are starting to emerge that give us a framework within which to rationalise the propensity for proteins to form oligomeric units, in terms of both specific functional advantages and generic facets of stability. As we begin to understand the evolutionary advantages of proteins adopting a particular oligomeric form, new applications of manipulating quaternary structure, for example in drug design and bionanotechnology are starting to emerge.

ACKNOWLEDGEMENTS

This work was funded, in part, by the Royal Society of New Zealand Marsden Fund. JAG thanks the University of Canterbury for funding and the Bio21 Institute, University of Melbourne, for hosting, a sabbatical leave.

REFERENCES

1. Eisenstein E, Schachman HK. Determining the roles of subunits in protein function. In: Creighton TE, ed. Protein Function. Oxford: IRL Press, 1989.
2. Perham RN. Self-assembly of biological macromolecules. Philos Trans R Soc Lond B Biol Sci 1975; 272(915):123-136.
3. Marianayagam NJ, Sunde M, Matthews JM. The power of two: protein dimerization in biology. Trends Biochem Sci 2004; 29(11):618-625.

4. Ali MH, Imperiali B. Protein oligomerization: How and why. Bioorg Med Chem 2005; 13(17):5013-5020.
5. Klotz IM, Darnall DW, Langerman NR. Quaternary structure of proteins. In: Hill RL, ed. The proteins. New York: Academic Press, 1975.
6. Beernink PT, Tolan DR. Disruption of the aldolase A tetramer into catalytically active monomers. Proc Natl Acad Sci USA 1996; 93(11):5374-5379.
7. Traut TW. Dissociation of enzyme oligomers: a mechanism for allosteric regulation. Crit Rev Biochem Mol Biol 1994; 29(2):125-163.
8. Monod J, Wyman J, Changeux JP. On the nature of allosteric transitions: a plausible model. J Mol Biol 1965:128-118.
9. Nooren IM, Thornton JM. Diversity of protein-protein interactions. EMBO J 2003; 22(14):3486-3492.
10. Goodsell DS, Olson AJ. Structural symmetry and protein function. Annu Rev Biophys Biomol Struct 2000:2905-153.
11. Ponstingl H, Thomas KB, Gorse D et al. Morphological aspects of oligomeric protein structures. Prog Biophys Mol Biol 2005; 89(1):9-35.
12. Papoian GA, Ulander J, Wolynes PG. Role of water mediated interactions in protein-protein recognition landscapes. J Am Chem Soc 2003; 125(30):9170-9178.
13. Ma B, Elkayam T, Wolfson H et al. Protein-protein interactions: structurally conserved residues distinguish between binding sites and exposed protein surfaces. Proc Natl Acad Sci, USA 2003; 100(10):5772-5777.
14. Ofran Y, Rost B. Analysing six types of protein-protein interfaces. J Mol Biol 2003; 325(2):377-387.
15. Jones S, Thornton JM. Principles of protein-protein interactions. Proc Natl Acad Sci, USA 1996; 93(1):13-20.
16. Jones S, Thornton JM. Analysis of protein-protein interaction sites using surface patches. J Mol Biol 1997; 272(1):121-132.
17. Jones S, Thornton JM. Protein-protein interactions: a review of protein dimer structures. Prog Biophys Mol Biol 1995; 63(1):31-65.
18. Lo Conte L, Chothia C, Janin J. The atomic structure of protein-protein recognition sites. J Mol Biol 1999; 285(5):2177-2198.
19. Nooren IM, Thornton JM. Structural characterisation and functional significance of transient protein-protein interactions. J Mol Biol 2003; 325(5):991-1018.
20. Jones S, Thornton JM. Prediction of protein-protein interaction sites using patch analysis. J Mol Biol 1997; 272(1):133-143.
21. Salwinski L, Eisenberg D. Computational methods of analysis of protein-protein interactions. Curr Opin Struct Biol 2003; 13(3):377-382.
22. Ofran Y, Rost B. Predicted protein-protein interaction sites from local sequence information. FEBS Lett 2003; 544(1-3):236-239.
23. Brooijmans N, Sharp KA, Kuntz ID. Stability of macromolecular complexes. Proteins 2002; 48(4):645-653.
24. Perutz MF. Science is not a quiet life: Unravelling the atomic mechanism of haemoglobin. Singapore: Imperial College Press, 1997.
25. Perugini MA, Griffin MDW, Smith BJ et al. Insight into the self-association of key enzymes from pathogenic species. Eur Biophys J 2005; 34(5):469-476.
26. Kim SY, Kim YW, Hegerl R et al. Novel type of enzyme multimerization enhances substrate affinity of oat beta-glucosidase. J Struct Biol 2005; 150(1):1-10.
27. Franke I, Meiss G, Blecher D et al. Genetic engineering, production and characterisation of monomeric variants of the dimeric Serratia marcescens endonuclease. FEBS Lett 1998; 425(3):517-522.
28. Mossing MC, Sauer RT. Stable, monomeric variants of λ Cro obtained by insertion of a designed β-hairpin sequence. Science 1990; 250(4988):1712-1715.
29. Borchert TV, Abagyan R, Jaenicke R et al. Design, creation and characterization of a stable, monomeric triosephosphate isomerase. Proc Natl Acad Sci, USA 1994; 91(4):1515-1518.
30. Jackson RM, Gelpi JL, Cortes A et al. Construction of a stable dimer of Bacillus stearothermophilus lactate dehydrogenase. Biochemistry 1992; 31(35):8307-8314.
31. Dobson CM. Getting out of shape. Nature 2002; 418(6899):729-730.
32. Nielsen L, Khurana R, Coats A et al. Effect of environmental factors on the kinetics of insulin fibril formation: elucidation of the molecular mechanism. Biochemistry 2001; 40(20):6036-6046.
33. Lashuel HA, Lai Z, Kelly JW. Characterization of the transthyretin acid denaturation pathways by analytical ultracentrifugation: implications for wild-type, V30M and L55P amyloid fibril formation. Biochemistry 1998; 37(51):17851-17864.
34. Foss TR, Kelker MS, Wiseman RL et al. Kinetic stabilization of the native state by protein engineering: Implications for inhibition of transthyretin amyloidogenesis. J Mol Biol 2005; 347(4):841-854.
35. Thorn DC, Meehan S, Sunde M et al. Amyloid fibril formation by bovine milk kappa-casein and its inhibition by the molecular chaperones alpha(s-) and beta-casein. Biochemistry 2005; 44(51):17027-17036.
36. Souillac PO, Uversky VN, Fink AL. Structural transformations of oligomeric intermediates in the fibrillation of the immunoglobulin light chain LEN. Biochemistry 2003; 42(26):8094-8104.

37. Raghu P, Reddy GB, Sivakumar B. Inhibition of transthyretin amyloid fibril formation by 2,4-dinitrophenol through tetramer stabilization. Arch Biochem Biophys 2002; 400(1):43-47.
38. Jenne DE, Denzel K, Blatzinger P et al. A new isoleucine substitution of Val-20 in transthyretin tetramers selectively impairs dimer-dimer contacts and causes systemic amyloidosis. Proc Natl Acad Sci USA 1996; 93(13):6302-6307.
39. Hammarstrom P, Sekijima Y, White JT et al. D18G transthyretin is monomeric, aggregation prone and not detectable in plasma and cerebrospinal fluid: a prescription for central nervous system amyloidosis? Biochemistry 2003; 42(22):6656-6663.
40. Clegg JS. Properties and metabolism of the aqueous cytoplasm and its boundaries. Am J Physiol 1984; 246(2 Pt 2):R133-151.
41. Makhatadze GI, Loladze VV, Ermolenko DN et al. Contribution of surface salt bridges to protein stability: guidelines for protein engineering. J Mol Biol 2003; 327(5):1135-1148.
42. Karshikoff A, Ladenstein R. Ion pairs and the thermotolerance of proteins from hyperthermophiles: a "traffic rule" for hot roads. Trends Biochem Sci 2001; 26(9):550-556.
43. Yip KS, Britton KL, Stillman TJ et al. Insights into the molecular basis of thermal stability from the analysis of ion-pair networks in the glutamate dehydrogenase family. Eur J Biochem 1998; 255(2):336-346.
44. Gerk LP, Leven O, Muller-Hill B. Strengthening the dimerisation interface of Lac repressor increases its thermostability by 40°C. J Mol Biol 2000; 299(3):805-812.
45. Vetriani C, Maeder DL, Tolliday N et al. Protein thermostability above 100°C: a key role for ionic interactions. Proc Natl Acad Sci USA 1998; 95(21):12300-12305.
46. Lebbink JH, Knapp S, van der Oost J et al. Engineering activity and stability of Thermotoga maritima glutamate dehydrogenase. II: construction of a 16-residue ion-pair network at the subunit interface. J Mol Biol 1999; 289(2):357-369.
47. Kou H, Pugh BF. Engineering dimer-stabilizing mutations in the TATA-binding protein. J Biol Chem 2004; 279(20):20966-20973.
48. Ahern TJ, Casal JI, Petsko GA et al. Control of oligomeric enzyme thermostability by protein engineering. Proc Natl Acad Sci USA 1987; 84(3):675-679.
49. Casal JI, Ahern TJ, Davenport RC et al. Subunit interface of triosephosphate isomerase: site-directed mutagenesis and characterization of the altered enzyme. Biochemistry 1987; 26(5):1258-1264.
50. Jones DH, McMillan AJ, Fersht AR et al. Reversible dissociation of dimeric tyrosyl-tRNA synthetase by mutagenesis at the subunit interface. Biochemistry 1985; 24(21):5852-5857.
51. Ward WH, Jones DH, Fersht AR. Protein engineering of homodimeric tyrosyl-tRNA synthetase to produce active heterodimers. J Biol Chem 1986; 261(21):9576-9578.
52. Ward WH, Jones DH, Fersht AR. Effects of engineering complementary charged residues into the hydrophobic subunit interface of tyrosyl-tRNA synthetase. Appendix: Kinetic analysis of dimeric enzymes that reversibly dissociate into inactive subunits. Biochemistry 1987; 26(13):4131-4138.
53. Carter P, Bedouelle H, Winter G. Construction of heterodimer tyrosyl-tRNA synthetase shows tRNATyr interacts with both subunits. Proc Natl Acad Sci USA 1986; 83(5):1189-1192.
54. Vora JP, Owens DR, Dolben J et al. Recombinant DNA derived monomeric insulin analogue: Comparison with soluble human insulin in normal subjects. Bmj 1988; 297(6658):1236-1239.
55. Brange J, Ribel U, Hansen JF et al. Monomeric insulins obtained by protein engineering and their medical implications. Nature 1988; 333(6174):679-682.
56. Hoffman A, Ziv E. Pharmacokinetic considerations of new insulin formulations and routes of administration. Clin Pharmacokinet 1997; 33(4):285-301.
57. DeFelippis MR, Chance RE, Frank BH. Insulin self-association and the relationship to pharmacokinetics and pharmacodynamics. Crit Rev Ther Drug Carrier Syst 2001; 18(2):201-264.
58. Miller MD, Krause KL. Identification of the Serratia endonuclease dimer: structural basis and implications for catalysis. Protein Sci 1996; 5(1):24-33.
59. Franke I, Meiss G, Pingoud A. On the advantage of being a dimer, a case study using the dimeric Serratia nuclease and the monomeric nuclease from Anabaena sp. strain PCC 7120. J Biol Chem 1999; 274(2):825-832.
60. LeFevre KR, Cordes MHJ. Retroevolution of lambda Cro toward a stable monomer. Proc Natl Acad Sci USA 2003; 100(5):2345-2350.
61. Borchert TV, Pratt K, Zeelen JP et al. Overexpression of trypanosomal triosephosphate isomerase in Escherichia coli and characterisation of a dimer-interface mutant. Eur J Biochem 1993; 211(3):703-710.
62. Borchert TV, Zeelen JP, Schliebs W et al. An interface point-mutation variant of triosephosphate isomerase is compactly folded and monomeric at low protein concentrations. FEBS Lett 1995; 367(3):315-318.
63. Borchert TV, Abagyan R, Kishan KVR et al. The crystal structure of an engineered monomeric triosephosphate isomerase, monoTIM: the correct modeling of an 8-residue loop. Structure 1993; 1(3):205-213.

64. Olivares-Illana V, Perez-Montfort R, Lopez-Calahorra F et al. Structural differences in triosephosphate isomerase from different species and discovery of a multitrypanosomatid inhibitor. Biochemistry 2006; 45(8):2556-2560.
65. Berg T. Modulation of protein-protein interactions with small organic molecules. Angew Chem-Int Edit 2003; 42(22):2462-2481.
66. Ryan DP, Matthews JM. Protein-protein interactions in human disease. Curr Opin Struct Biol 2005; 15(4):441-446.
67. Arkin MR, Wells JA. Small-molecule inhibitors of protein-protein interactions: Progressing towards the dream. Nat Rev Drug Discov 2004; 3(4):301-317.
68. Schliebs W, Thanki N, Jaenicke R et al. A double mutation at the tip of the dimer interface loop of triosephosphate isomerase generates active monomers with reduced stability. Biochemistry 1997; 36(32):9655-9662.
69. Borchert TV, Kishan KV, Zeelen JP et al. Three new crystal structures of point mutation variants of monoTIM: conformational flexibility of loop-1, loop-4 and loop-8. Structure 1995; 3(7):669-679.
70. Mainfroid V, Terpstra P, Beauregard M et al. Three hTIM mutants that provide new insights on why TIM is a dimer. J Mol Biol 1996; 257(2):441-456.
71. Mainfroid V, Mande SC, Hol WG et al. Stabilization of human triosephosphate isomerase by improvement of the stability of individual α-helices in dimeric as well as monomeric forms of the protein. Biochemistry 1996; 35(13):4110-4117.
72. McKenzie AN, Ely B, Sanderson CJ. Mutated interleukin-5 monomers are biologically inactive. Mol Immunol 1991; 28(1-2):155-158.
73. Dickason RR, Huston DP. Creation of a biologically active interleukin-5 monomer. Nature 1996; 379(6566):652-655.
74. Sano T, Vajda S, Smith CL et al. Engineering subunit association of multisubunit proteins: a dimeric streptavidin. Proc Natl Acad Sci USA 1997; 94(12):6153-6158.
75. Pazy Y, Eisenberg-Domovich Y, Laitinen OH et al. Dimer-tetramer transition between solution and crystalline states of streptavidin and avidin mutants. J Bacteriol 2003; 185(14):4050-4056.
76. Laitinen OH, Nordlund HR, Hytonen VP et al. Rational design of an active avidin monomer. J Biol Chem 2003; 278(6):4010-4014.
77. Velichko IS, Mikalahti K, Kasho VN et al. Trimeric inorganic pyrophosphatase of Escherichia coli obtained by directed mutagenesis. Biochemistry 1998; 37(2):734-740.
78. Salminen A, Efimova IS, Parfenyev AN et al. Reciprocal effects of substitutions at the subunit interfaces in hexameric pyrophosphatase of Escherichia coli. Dimeric and monomeric forms of the enzyme. J Biol Chem 1999; 274(48):33898-33904.
79. Fritsche P, Alves J. A monomeric mutant of restriction endonuclease EcoR I nicks DNA without sequence specificity. Biol Chem 2004; 385(10):975-985.
80. Zaremba M, Sasnauskas G, Urbanke C et al. Conversion of the tetrameric restriction endonuclease Bse634I into a dimer: Oligomeric structure-stability-function correlations. J Mol Biol 2005; 348(2):459-478.
81. Banci L, Benedetto M, Bertini I et al. Solution structure of reduced monomeric Q133M2 copper, zinc superoxide dismutase (SOD). Why is SOD a dimeric enzyme? Biochemistry 1998; 37(34):11780-11791.
82. Banci L, Bertini I, Chiu CY et al. Synthesis and characterization of a monomeric mutant Cu/Zn superoxide dismutase with partially reconstituted enzymic activity. Eur J Biochem 1995; 234(3):855-860.
83. Porvari KS, Herrala AM, Kurkela RM et al. Site-directed mutagenesis of prostatic acid phosphatase. Catalytically important aspartic acid 258, substrate specificity and oligomerization. J Biol Chem 1994; 269(36):22642-22646.
84. Breiter DR, Resnik E, Banaszak LJ. Engineering the quaternary structure of an enzyme: construction and analysis of a monomeric form of malate dehydrogenase from Escherichia coli. Protein Sci 1994; 3(11):2023-2032.
85. Beernink PT, Tolan DR. Subunit interface mutants of rabbit muscle aldolase form active dimers. Protein Sci 1994; 3(9):1383-1391.
86. Beernink PT, Tolan DR. Disruption of the aldolase A tetramer into catalytically active monomers. Proc Natl Acad Sci USA 1996; 93(11):5374-5379.
87. Bailey DL, Fraser ME, Bridger WA et al. A dimeric form of Escherichia coli succinyl-CoA synthetase produced by site-directed mutagenesis. J Mol Biol 1999; 285(4):1655-1666.
88. Griffin MDW, Dobson RCJ, Pearce FG et al. Evolution of quaternary structure in a homotetrameric enzyme. J Mol Biol 2008; 380(4): 691-703.

CHAPTER 6

OLIGONUCLEOTIDE BINDING PROTEINS

The Occurrence of Dimer and Multimer Formation

Jackie Wilce,* Julian Vivian and Matthew Wilce

Department of Biochemistry and Molecular Biology, and School of Biomedical Sciences, Monash Univerity, Melbourne, Victoria, Australia

**Corresponding Author: Jackie Wilce—Email: jackie.wilce@med.monash.edu.au*

Abstract: Protein dimers and multimers are often employed by nature for DNA and RNA handling and formation of specific, high-affinity protein-oligonucleotide complexes. The repeating structure of dsDNA lends itself to recognition by multimeric protein complexes that can assemble about the helical structure. In the cases of both DNA and RNA, specific recognition of nucleotide sequences can be achieved by multidomain proteins or protein multimers. Furthermore large multimeric assemblies are utilised for the stable formation of structures such as rings and filaments. Also, the assembly of multimeric structures by interchangeable subunits can add layers of regulation and increase functional complexity. Thus there appear to be many advantages to oligonucleotide interactions that are conferred by dimerisation or multimerisation.

INTRODUCTION

Large proteins are required for various biological functions. This can be achieved by a single polypeptide comprising several domains, or a multimeric assembly of small proteins. In most cases nature appears to favour multimeric assembly. Large multimeric assemblies are utilised for the stable formation of structures such as rings and filaments. Also, the assembly of multimeric structures by interchangeable subunits can add layers of regulation and increase functional complexity. The average size of a polypeptide is ~30 kDa. This may be a control mechanism to prevent error in translation synthesis as noted by Goodsell et al.[1] An estimate of prokaryotic translational error resulting in termination of the product is ~3×10^{-4} per codon, suggesting that a 500 amino acid protein (~60 kDa) would be prematurely terminated 1 in 7 times. A 3000 amino acid protein would rarely

Protein Dimerization and Oligomerization in Biology, edited by Jacqueline M. Matthews.

be fully translated.[2] Thus the utilisation of smaller protein building blocks is an elegant solution for forming large protein surfaces.

Protein dimers and multimers are highly prevalent in their role as oligonucleotide binding partners. In their requirement to make both specific and nonspecific interactions with DNA and RNA, dimers, multimers, rings and extended multimers form in a spectacular array of quaternary designs. The adoption of multimeric proteins may have simply arisen generally for the economy of genetic material and to avoid translational error, but in this chapter we identify four circumstances in which dimer and multimer formation appear to provide an optimum solution for forming interactions with oligonucleotides. Firstly, the multimer is better able than the monomer to make high affinity specific interactions with linear tracts of RNA or DNA. Secondly, oligonucleotide interactions can readily be regulated by manipulation of a dimeric protein. Furthermore, the presence of multiple domains provides an efficient way to initiate a response to a variety of signals using a relatively limited number of polypeptides. Thirdly, due to the repeating nature of oligonucleotides, the handling of long tracts is best facilitated by a multimeric protein. And fourthly, where a highly stable ring or scaffold structure is required for oligonucleotide processing, this is most readily achieved by the assembly of a multimeric structure.

These themes are elaborated on in four sections in this chapter. They are illustrated by nine selected cases that are intended to showcase a spectrum of biological processes involving multimeric protein/oligonucleotide interactions. The selected examples are far from exhaustive and vary in their prevalence in biological systems. The cases are restricted to examples where biological function is well characterized and structural information is available. This has resulted in our showcasing a greater number of DNA based systems than RNA based systems—but this does not necessarily reflect a lesser role of protein multimers in RNA biology. Finally, the occurrence of dimers and multimers in binding oligonucleotides is contrasted with some examples where nature has chosen to utilize monomeric proteins to the same end.

SPECIFIC AND HIGH-AFFINITY PROTEIN-OLIGONUCLEOTIDE INTERACTIONS

The recognition of DNA and RNA sequences by proteins is fundamental to cellular processes—from the initiation of transcription via the recognition of promotor sequences and the regulation of DNA replication to the recognition of functional RNA. The interaction with an average sized monomeric protein along an elongated stretch of DNA or RNA is intrinsically limited. Thus, proteins that form specific interactions with oligonucleotide sequences often bind as dimers or multimers to confer extra specificity and affinity. A common mode of interaction with dsDNA, for example, is via a recognition α-helix inserted at a major groove where contacts to unique functionalities are made.[3] In this arrangement, no more than five consecutive base pairs can contribute to the protein/DNA interface.[4] Both sequence specificity and affinity are thus limited unless further contacts are made. An efficient solution to the problem is multimer formation in which greater specificity and affinity are conferred due to multivalent binding.[5]

Two examples of the way in which multimer formation has been adopted by proteins for specific recognition and enhanced affinity for dsDNA are described below. They are the Type II endonucleases and the nuclear receptor transcription factors which both act as dimers. It is of interest to note that in the case of specific recognition of RNA, multivalent binding has also evolved. Of the well characterized proteins that function via specific recognition of RNA sequences, however, multidomain structures tend to have been adopted rather than dimers or multimers. The covalently linked domains can each contribute to either RNA specificity, affinity or both.[6,7] The presence of multiple domains appears to equally well confer the necessary specificity and affinity for target RNA sequences.

CASE 1—Type II Endonucleases Are Functional Homodimers Reflecting the 2-Fold Symmetry of Their Target DNA Sequence

Type II restriction endonucleases are a subset of site-specific DNA-cutting enzymes found in bacteria and archaea. They act as innate immune systems to protect the host from invasion by foreign DNA such as viral DNA.[8] They cleave both strands of dsDNA with remarkable specificity. Double-strand breaks are generally lethal so the host protects its own genome from cleavage by pairing each Type II restriction enzyme with a cognate methyltransferase that methylates the N4 or C5 of cytosine or the N6 of adenine of target sites. These modifications ensure the restriction enzyme distinguishes host DNA.[9] The target sequences for Type II restriction endonucleases are typically 4-8 base pair palindromes or palindromes separated by a stretch of nonconserved bases. Cleavage occurs at equivalent sites on each strand of the dsDNA in or near the target sequence.

Type II restriction endonucleases typically form functional homodimers, their structural and functional symmetry reflecting the 2-fold symmetry of their target sequence.[9] i.e., the two monomers simultaneously bind to the dsDNA and cleave a strand each. Typical examples of Type II restriction endonucleases are *Eco*RV[10] and *Eco*RI[11] (Fig. 1A). Both proteins form symmetrical homodimers, with *Eco*RV recognising the palindrome GATATC and *Eco*RI recognising the palindrome GAATTC. Each monomer recognizes one-half of the palindrome, which is only 2-4 base pairs long. Thus as well as facilitating double strand cleavage, dimer formation requires the recognition of 4-8 base pairs, ensuring greater stringency in target recognition.

One of the very few exceptions to the dimeric assembly of Type II restriction endonucleases is *Mva*I.[12] This enzyme is a monomer that recognizes the pseudo-palindromic sequence CCWGG, where W is A or T. The structural difference of *Mva*I is reflected in its altered mechanism. *Mva*I doesn't act symmetrically like the other Type II restriction enzymes. Instead, it binds the dsDNA in an asymmetric manner. Also, instead of cutting the two strands of dsDNA simultaneously, it first cleaves one strand then moves around the DNA and cleaves the second strand.[12] Clearly, whereas a monomeric enzyme is suited to cleaving a single strand of DNA, the formation of a dimer is ideal for the recognition and simultaneous cleavage of both strands of dsDNA.

CASE 2—Nuclear-Receptor Dimerisation Facilitates Target Specificity

The nuclear receptor family represents the largest family of transcriptional regulators in multicellular organisms. They are responsible for initiating transcriptional responses to hormones such as the sex steroids (estrogen, progesterone and testosterone) as well as many other small molecule ligands.[13] Upon ligand activation they undergo the formation of homo- or hetero dimers and are able to bind directly to their target DNA response elements to activate or repress transcription (although some nuclear receptors have been shown to act as monomers.[14] Ligand binding results in a structural reconfiguration of

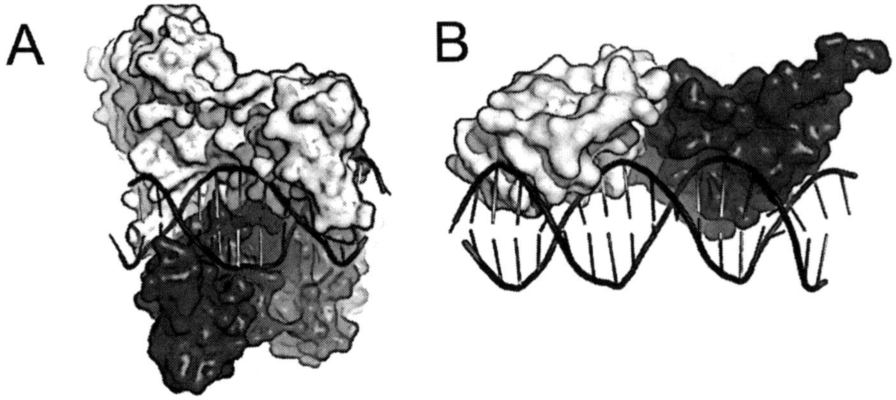

Figure 1. Specific and high-affinity protein-oligonucleotide interactions. A) Schematic of the dimeric restriction enzyme EcoRI bound to duplex DNA: PDBID 1ERI. B) Schematic of the dimeric DNA-binding domain of the human estrogen receptor in complex with duplex DNA: PDBID 1HCQ. The molecular surfaces are shown in different shades to indicate the different subunits. The DNA is shown in cartoon representation.

the nuclear receptor that initiates the recruitment of coactivators or corepressors which impact on chromatin remodelling and the final transcriptional response.

The architecture of nuclear receptors is conserved, with all possessing an N-terminal domain involved in gene activation, a double zinc-finger DNA-binding domain (DBD), a connecting hinge region and a discrete ligand binding domain (LBD). Dimerisation occurs both between LBDs (in a ligand dependant manner) and through the DBD to form a structure that binds to response element sequences, with each monomer recognising a hexameric half-site. The mode of interaction between DBD dimers and DNA is illustrated by structures of the glucocorticoid receptor (GR) and estrogen receptor (ER) bound to their cognate DNA sites (Fig. 1B). Binding to target DNA is conferred via the interaction of the α-helix of DBD zinc-finger I with a hexameric half site. Zinc finger II is oriented to form a dimerisation interface with the other DBD. This interface positions the two DBDs with the correct spacing for binding to the two-hexameric half sites of the response element.

The binding of DBDs is strongly cooperative and important for the overall affinity of the nuclear receptor/DNA complex. Dimerisation also facilitates specific recognition of response elements by requiring a bivalent interaction with two hexameric half site sequences that are arranged with the correct orientation and spacing. Whereas there is little discrimination by a nuclear receptor monomer in binding to a hexameric-half site, the combined sequences and geometry of the two half-sites facilitates the specific recognition by a nuclear receptor complex.[13] For example, the GR and other steroid receptors binds to a target response element containing symmetric (palindromic) repeats of the hexameric half site sequence 5′-AGAACA-3′. Here a three base pair separation between hexameric half sites is required.[15] The ER also binds to symmetric hexameric half sites, but recognizes a different sequence: 5′-AGGTCA-3′. In contrast retinoid X receptor (RXR) homo- and heterodimers bind to direct repeats of the 5′-AGGTCA-3′ sequence. Here a spacing of 1-5 base pairs between the repeats dictates which of the various RXR heterodimers will bind. Thus dimerisation is an important feature of DNA selectivity.

REGULATION OF PROTEIN BINDING TO OLIGONUCLEOTIDE

Dimer or multimer formation by proteins is well known to facilitate the allosteric regulation of protein function. This is also the case for oligonucleotide binding proteins where allosteric or cooperative interactions between monomers can determine the binding affinity for the target oligonucleotide. The control can occur via several different molecular mechanisms: (1) via structural changes imparted to the protein multimer by a ligand, (2) by the variable formation of heterodimers or (3) by cooperative interaction between proteins upon their binding to the target oligonucleotide.

One of the largest and best studied classes of DNA-binding proteins are the transcriptional regulators. These molecules bind to specific sites upstream of the gene being regulated and initiate or repress transcription via interactions with RNA polII, general transcription factors and other cofactors.[16] They interact with duplex DNA via well known motifs including helix-turn-helix, helix-loop helix, zinc-finger and basic region leucine zipper (bZIP) motifs. These motifs are all relatively small structural motifs that form specific interactions with the major groove of dsDNA via an α-helix. They can occur as homo or heterodimers which facilitates their regulation. Examples are illustrated below. See also Chapter 7 by Funnell and Crossley.

CASE 3—TrpR Is an Allosterically Regulated Dimeric Transcriptional Repressor

The tryptophan repressor (TrpR) is a dimeric transcriptional regulator found in enteric bacteria. When activated by L-tryptophan it binds to specific DNA sequences at the operator site of the operons responsible for biosynthesis of L-tryptophan. When two molecules of L-tryptophan are bound to TrpR the repressor is able to strongly bind the operator thereby preventing transcription of L-tryptophan biosynthetic genes. In the absence of L-tryptophan the apo-repressor is unable to bind the operator and L-tryptophan synthesis is upregulated. In this way TrpR regulates the levels of available L-tryptophan in the cell (reviewed in ref. 17).

TrpR is a homodimer comprising two helix-turn-helix DNA binding sites (Fig. 2A). The dimerisation core of the protein comprises four helices from each monomer that interlock to form a remarkably tightly locked complex. Conversely, the helix-turn-helix DNA binding domains flanking the dimerisation core appear to be relatively mobile. The two L-tryptophan molecules (one per subunit) bind between the helix-turn-helix motif of one subunit and a helix of the partner subunit.[18]

L-tryptophan acts as an allosteric activator of the repressor by increasing and locking in the spacing between the DNA binding domains. The distance observed in the crystal structure of the apo-TrpR is 25 Å. The binding of two L-tryptophan molecules results in conformational changes to the protein that increase this DNA binding domain distance to 34 Å (the distance between successive B-form DNA major grooves) allowing the repressor to recognize its binding site.[19,20] In this way TrpR acts as an allosterically controlled molecular calliper.

CASE 4—Polar Fork Arrest in *B. subtilis* Requires the Formation of a Dimer of Dimers

An unusual but striking example of binding cooperativity is found in the regulation of polar DNA replication fork arrest by the replication termination protein (RTP) of *B. subtilis*. Here the functional complex requires the formation of a dimer of RTP dimers at termination sequences in the chromosome known as the *Ter* sites. The RTP: *Ter* complex is able to halt or pause the approaching replicative helicase when it approaches from one direction, but not the other.

The mechanism underlying polar fork arrest by the RTP: *Ter* complex relies upon the cooperative binding between two adjacent RTP dimers that bind to a high affinity and low affinity site within the *Ter* sequence (Fig. 2B). RTP binds to DNA via a "winged-helix"[21] motif and it is thought that the wings facilitate the binding cooperativity observed.[22,23] Replication fork arrest only occurs when the high affinity site is proximal to the approaching replicative helicase. When the replicative helicase approaches the low affinity site it is able to knock off the bound RTP and pass through the terminator complex.[24] The complex is cooperative in that the low affinity site will not be bound by RTP without prior filling of the high affinity site. Furthermore, the binding of the high affinity site alone is insufficient to halt the replicative helicase.

Interestingly, cooperative formation of a dimer is not the only way nature has managed to effect polar fork arrest. A polar replication terminator system has also been characterized in *E. coli* that bears no sequence or structural similarity to RTP.[25] Unlike RTP, which has evolved as a dimer of a common DNA binding motif, the *E. coli* equivalent, Termination utilisation substance (Tus), is a protein of unique fold and operates as a monomer. As a monomer, Tus is not able to exploit cooperativity and arrests the helicase by a different mechanism. The Tus: *Ter* complex blocks the action of the replicative helicase by virtue of its intrinsic asymmetry.[26] The translocation of the helicase along the DNA forms a stable locked complex only at the nonpermissive end of the Tus: *Ter* complex via the shifting of the C(6) nucleotide into a cytosine specific binding site on Tus.[26] This renders the Tus: *Ter* complex a polar nonspecific block to helicase translocation along the DNA.

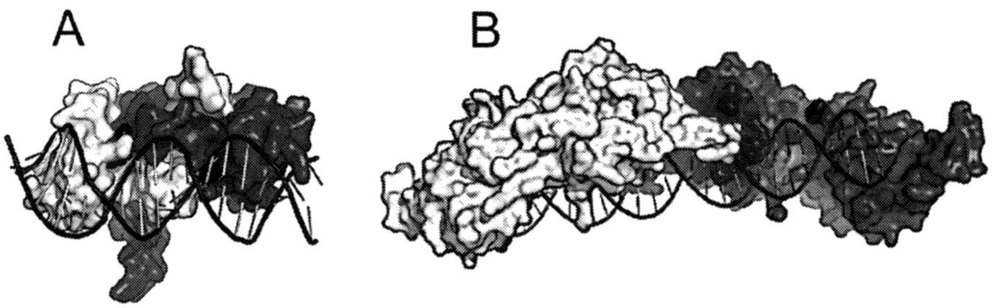

Figure 2. Regulation of oligonucleotide binding. A) Schematic of the dimeric TrpR bound to duplex DNA: PDBID 1TRO. B) Representation of a dimer of dimers of the replication termination protein bound to dsDNA. (Model based on coordinates PDBID 1F4K). The molecular surfaces are shown in different shades to indicate the different subunits. The DNA is shown in cartoon representation.

MODIFICATION OF OLIGONUCLEOTIDE ARCHITECTURE BY MULTIMERIC PROTEINS

Many cellular processes require the manipulation of long oligonucleotide tracts at either specific or nonspecific sequences. Where handling of a long tract is required, the repeating nature of the oligonucleotide is often reflected in the multimeric structure of the protein binding partner. Following are three examples of protein multimers that form spectacular complexes with DNA and RNA in diverse biological processes.

CASE 5—Packaging of the Eukaryotic Chromosome by Histones

The ultimate example of modified oligonucleotide architecture is almost certainly the packaging of the eukaryotic chromosome into chromatin. Chromatin is a fibrous superstructure composed of nucleosome core particles (NCP). Each NCP contains 146 bp of dsDNA wrapped in two superhelical turns around an octamer of histones.[27] The histone octamer is formed by the assembly of two each of class H2A, H2B, H3 and H4 core histones or variants of these.[28] Higher order structure is formed by the arrangement of NCPs into a fibre of 30 nm diameter via histone tails and the linker histone (H1). Several additional levels of folding are then required to pack the entire length of DNA (2 m in humans) into the nucleus.[29]

The core histones are relatively similar in structure and are highly conserved through evolution (Fig. 3A). DNA binding and octamer formation is conferred via a "helix-turn-helix-turn-helix" motif. Interactions with the DNA are nonsequence specific and primary contacts are made by the highly basic N-termini of the α2-helices of all four types of histone and the N-termini of α1-helices of histones H2B, H3 and H4 with the phosphate groups of the DNA backbone. The helix dipoles lock these phosphates into place. Other interactions include contacts between the histone loop regions and the DNA backbone. The histones also possess long 'tails' on one end of the amino acid structure through which chromatin structure is regulated.

Again, nature has utilized a multimer for an important task, DNA packaging. This appears to serve several functions. Firstly, the core histone structure is clearly optimized to make strong nonspecific interactions with DNA, capable of overwinding dsDNA necessary for tight compaction. The repeating nature of dsDNA is hence reflected in the repeated use of this optimized histone module. Secondly, the octameric nature of the histone facilitates the modulation of chromatin flexibility that is required during processes such as DNA replication, transcription and repair. The modular nature of the histone octamer permits histone variants to be readily incorporated into the chromatin structure affecting its dynamic behaviour.[28]

CASE 6—The TRAP Multimer Impacts on mRNA Secondary Structures Involved in Transcription and Translation

In another case of the regulation of tryptophan biosynthetic genes (see CASE 3 above), a multimeric oligonucleotide binding protein is involved. Unlike *E. coli*, *Bacillus* species possess no *trp*-repressor. Instead, control of gene expression occurs posttranscriptionally, at the level of mRNA. The Trp RNA-binding attenuation protein (TRAP) of *Bacilli* binds at a specific sequence in the leader sequence of the tryptophan biosynthetic mRNA

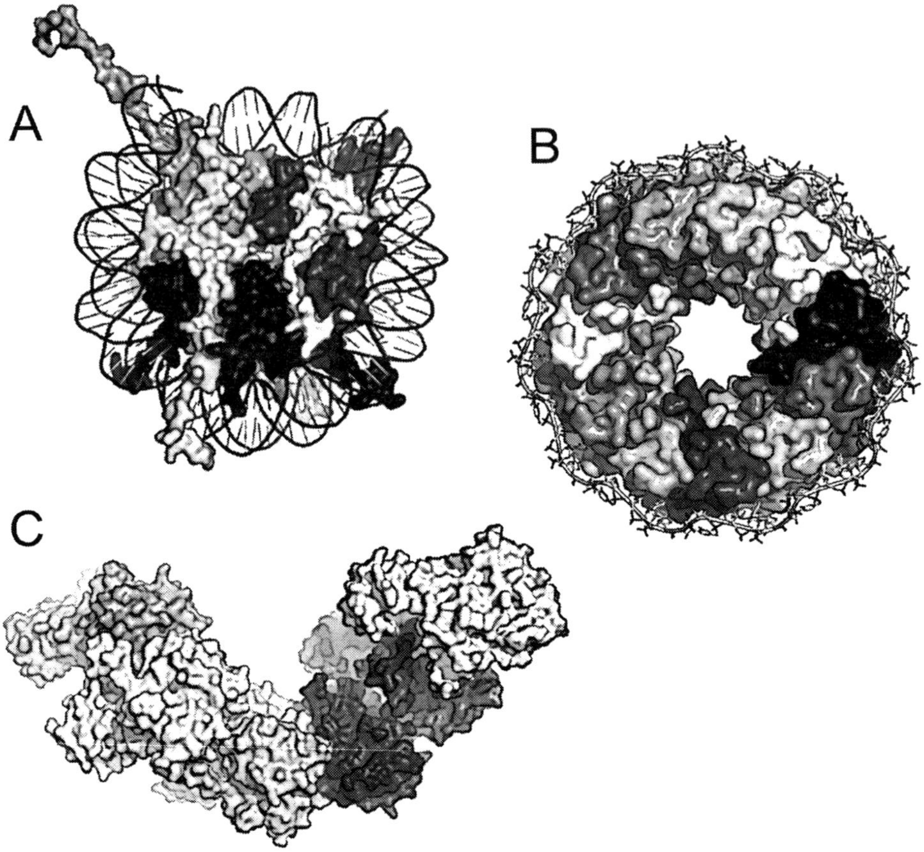

Figure 3. Modification of oligonucleotide architecture by multimeric proteins. A) Schematic of the core histone octomer in complex with dsDNA: PDBID 1AOI. B) Trp RNA binding attenuation protein (TRAP) oligomer with DNA: PDBID 1C9S. The DNA can be seen around the periphery of the molecule C) Representation of the Rad51 filament based upon coordinates from PDBID 1SZP. The molecular surfaces are shown in different shades to indicate the different subunits. The DNA is shown in cartoon representation.

transcript when bound by L-tryptophan and as it is being actively transcribed. This results in the formation of a large protein/RNA complex which interferes with adjacent mRNA secondary structures which are essential for the continuation of transcription.[30]

The RNA target for TRAP involves eleven trinucleotide repeats, each of which is recognized by a TRAP subunit arranged in a large ring[31] (Fig. 3B). This interaction forces the mRNA to adopt a dramatically different architecture from what it would otherwise adopt, at not only the immediate site but also the adjacent sequences. In particular an "anti-terminator" stem loop is disrupted and a "terminator" stem loop is formed. This interferes with the continuation of transcription. Furthermore, even if the mRNA is fully transcribed, the complex of TRAP at the leader region of the transcript interferes with translation. This is due to the impact of the structure on the downstream ribosome binding site which inhibits ribosome binding. The TRAP system is thus an elegant example of the way in which oligonucleotide architecture can be manipulated by interaction with a protein multimer.

CASE 7—The Recombinase Polymer Provides an Extended Platform for the Handling of DNA

Homologous recombination is an important mechanism in dsDNA handling, conserved from bacteria through to humans, which performs two seemingly opposite functions. Firstly, it maintains the stability of the genome through the repair of double-strand breaks that are the result of DNA damage or that are induced during meiosis. Secondly, it generates genetic diversity through chromosome crossover. Homologous recombination uses stretches of homologous dsDNA as a template to pair and exchange with broken DNA to produce two intact dsDNA molecules.[32-34]

The process of homologous recombination involves several enzymes. The doublestranded breaks are recognized and processed by specific helicases that unwind the dsDNA and specific exonucleases pare back the 5′-strand producing long 3′ ssDNA tails.[35] Recombinases then catalyse the pairing and exchanging of homologous strands. In *E. coli* the predominant recombinase is RecA.[36] The eukaryotic equivalents to RecA are Rad51[37] that is involved in various forms of double-strand break repair and DMC1[38] that is meiosis specific. These enzymes form extended polymers on the ssDNA tails to produce a nucleo-protein filament. The formation of the recombinase polymers occurs in the 5′ to 3′ direction, forming a right-handed nucleo-protein spiral that removes the secondary structure of the ssDNA. The nucleo-protein filament then aligns itself with the homologous chromosome at the site where the ssDNA tail can be paired. This proceeds by the nucleo-protein filament first binding to the dsDNA nonspecifically to form a three-stranded DNA complex and then moving to a region of sequence homology.[39] The "invading" ssDNA strand pairs with the homologous dsDNA and is exchanged with one strand of the homologous dsDNA to form hetero-dsDNA. DNA polymerase generates new dsDNA, primed by the "invading" strand and complimentary to the intact strand. This creates a mobile region (known as a Holliday junction) at which the two dsDNA molecules are interlinked.[32,36] The Holliday junctions are resolved by specific protein complexes into two recombined dsDNA molecules.[40]

Structurally the *E. coli* and eukaryotic recombinases are very similar and also share homology with hexameric helicases.[36] In the formation of the nucleo-protein filament each RecA monomer binds 3 nucleotides of ssDNA and, like the hexameric helicases, there are six monomers per turn (Fig. 3C). The prototypic RecA monomer is a 38 kDa protein comprising three domains. The N-terminal domain comprises 30 amino acids and forms contacts with neighbouring subunits in the filament. The core domain consists of 230 residues and contains the ssDNA binding site. It also contains the ATP binding and hydrolysis site that is in contact with an adjacent subunit. The recombinases utilize the hydrolysis of ATP to drive the conformational changes to the nucleo-protein filament required for homologous strand-exchange. The C-terminal domain has a surface on the outside of the filament to which dsDNA binds.[36]

The recombinases provide an extended platform for the association of ssDNA with homologous dsDNA, facilitate base-pairing to the complementary sequence and coordinate the strand-exchange. The handling of a long linear ssDNA is required and this is best managed by a multimer that can associate with the repeating nucleotide structure of varying lengths. Furthermore the association of the nucleo-protein filament with lengths of chromosomal dsDNA must occur with great efficiency in order to locate the homologous region within the genome. Exonucleolytic processing of double-strand breaks can generate over 1000 base pairs of ssDNA. The search for homology to a

1000 base pair stretch in a human cell has been likened to searching for a linear 20 cm expanse in 2500 km.[32] The recombinase polymer is also able to provide the coordinated conformational change required for strand exchange. The details of this remarkable process are still being elucidated.

MULTIMERIC-RINGS IN OLIGONUCLEOTIDE PROCESSING

Many proteins involved in DNA and RNA processing have adopted a ring shaped structure by the assembly of homo or heteromeric multimers. The ring possesses a structural stability that is required for highly stringent processes and the ability to encircle the target oligonucleotide facilitating movement along its length. Rings that encircle DNA have to be large enough to fold to maintain a central cavity of 20-34Å and, for some, conduct catalytic activities. Most tend to be hexameric or have pseudo-hexameric symmetry and it has been suggested that the hexameric arrangement of subunits is a compromise between folding requirements and minimizing size.[41] The formation of a ring structure via multimerisation (rather than the assembly of structurally unrelated proteins) reflects the repeating nature of the oligonuclotide with which each subunit may interact. Furthermore, the requirement for ring assembly provides a mechanism for the regulation of the oligonucleotide processing. Multimeric rings are utilized in DNA replication, repair and recombination as well as RNA handling. Some examples are described below.

CASE 8—Sm and Lsm Heteromeric Heptamers in RNA Processing

The Sm and Lsm (Sm-like) families are multimeric proteins that play integral roles in the processing of mRNA. One of the best characterized functions of the Sm proteins is as a component of the spliceosome in eukaryotes. Spliceosomes remove the introns out of preRNA and seal the exon ends together. They are made up of one of several species of small nuclear RNAs (snRNAs; U1, U2, U4, U5 and U6) bound to an extremely stable heptameric Sm core to form a set of small nuclear ribonucleoproteins (snRNPs). Splicing occurs when the series of snRNPs interact with the pre-mRNA in a particular order and at specific sites in the pre-mRNA. That is, U1 snRNP binds to the 5′-end of the intron and U2 snRNP binds close to the 3′-end. This event is followed by binding by U4/U6 snRNPs and, finally, U5 snRNP, which helps to hold the exons together.

The Sm proteins do not bind directly to mRNA. Specific binding to the splice sites in the pre-mRNA is dictated by base pairing with the bound snRNA. The role of the Sm heptameric ring is to provide a stable scaffold for tethering the snRNA at its consensus Sm-binding sequence (GAU_4GA). The Sm heptamers are heteromeric. Each spliceosomal snRNP is each made up of the Sm proteins B or B', D1, D2, D3, E, F and G, except for the U6 snRNP. The Sm heptamer forms a highly cationic pore at the centre, which is the site for snRNA binding (Fig. 4A).

A clue as to why the spliceosomal Sm protein has evolved to operate as a multimer may lie in the requirement for a highly stable protein-RNA complex that is assembled with high fidelity.[42] In vitro, snRNP complexes form spontaneously with uridine-rich RNA. In vivo, however, this is not allowed to happen. Instead they are assembled in the cytoplasm via a highly stringent mechanism involving the SMN (survival of motor neurons) complex. The SMN complex distinguishes snRNAs from other uridine-rich RNA sequences and identifies the RNA-binding Sm proteins, bringing them together

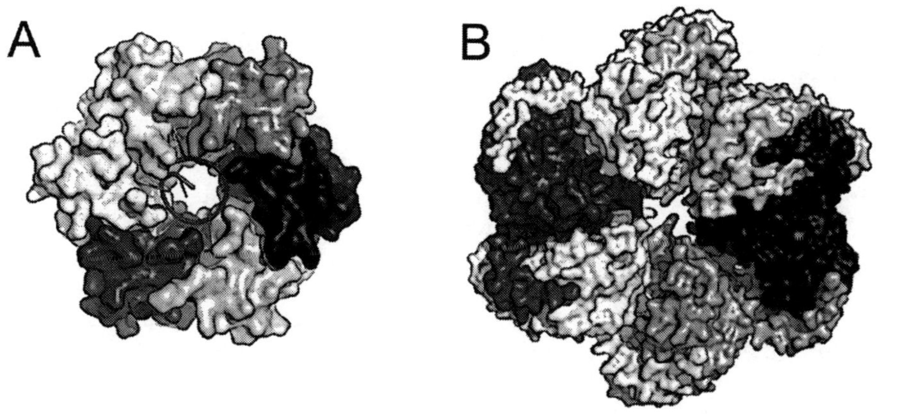

Figure 4. Multimeric-rings in oligonucleotide processing. A) Schematic of the Sm-like Hfq protein from *S. aureus* in complex with RNA: PDBID 1KQ2. B) Structure of a classical ring helicase. No crystal structure with DNA are available: PDBID 1E0J. The molecular surfaces are shown in different shades to indicate the different subunits. The DNA is shown in cartoon representation.

around the snRNA. The final structure is very stable—there is no evidence that there is any turnover of snRNP complexes. Thus any improper association with other RNA sequences that could lead to improper splicing is avoided.

In addition to Sm proteins, several Sm-like (Lsm) proteins have been described in eukaryotes. These are able to form heteromeric rings and are involved in a diverse range of functions including pre-mRNA splicing, processing of nuclear mRNAs and in decay of mRNAs.[43] Their compositions vary, with different combinations of the ~16 Lsm proteins identified to date, defining the Lsm heptamer function. For example, Lsm2-8 (composed of seven Lsm proteins numbered 2-8) is localized to the nucleus where it carries out preRNA splicing and nuclear mRNA processing functions. Lsm1-7, alternatively, is localized to processing bodies (P-bodies) in the cytoplasm and is required for the decapping step of mRNA decay after deadenylation. The heteromeric compositions of the Lsm complexes thus appear to facilitate a versatility of function, including their ability to recognize different RNA sequences.

CASE 9—Sliding-Clamps and Hexameric Helicases

Many proteins involved in the metabolism of large stretches of DNA and RNA execute their function whilst translocating along single-stranded or double-stranded oligonucleotides. Of these proteins, a large number have evolved into structures that encircle the oligonucleotide. Of these, the predominant morphology is the formation of multimeric rings.

One well-characterized family of proteins that encircle dsDNA is the "sliding-clamp" family involved in DNA replication. In this process they are loaded onto the newly synthesized dsDNA by a clamp-loader complex.[44] Once loaded the sliding-clamp is bound by the replicative DNA polymerase and acts as a mobile collar to tether the DNA polymerase to the DNA template. In so doing, the sliding greatly increases the processivity of the DNA polymerase. That is, the polymerase can catalyse a greater amount of dsDNA before dissociating from the template. In the absence of the sliding-clamp the polymerase can only process a few nucleotides. This increases 1000-fold with the addition of the sliding-clamp.[45]

Sliding-clamps all form multimeric rings with pseudo-sixfold symmetry (Fig. 4B). The *E. coli* β-clamp is a dimer of three homologous domains.[46] The human equivalent PCNA (proliferating cell nuclear antigen) is a trimer of two domains.[47] The domains of these proteins appear to have fused in alternative arrangements during gene duplication. Structurally the β-clamp and PCNA proteins are very similar. They form planar rings with a central cavity of 35 Å, wide enough to enclose dsDNA. They are thought to interact with the DNA solely through bridging water molecules that act as a lubricant to allow the clamps to slide along the dsDNA with minimal hindrance.

Hexameric helicases are another family of oligonucleotide processing proteins that utilize a central cavity to associate tightly with DNA whilst maintaining the ability to translocate along the oligonucleotide. They have been shown to play a role in DNA replication, recombination and transcription.[48] Unlike the sliding-clamps, helicases drive their own translocation by the hydrolysis of NTP (nucleoside triphosphate). This translocation is intimately linked with their function, which is to unwind dsDNA or dsRNA to produce single-stranded lengths of oligonucleotide required for various metabolic processes. For example, in *E. coli* the ssDNA templates required for DNA replication are produced by the hexameric helicase DnaB[49] and branch migration during homologous recombination is driven by the hexameric helicase RuvB.[50]

Hexameric helicases predominantly assemble as homohexamers with ssDNA or ssRNA binding in the central cavity.[48] The other stand of the duplex is thought to be excluded from the central cavity, remaining on the outside of the ring. Each subunit of the hexamer contains an NTPase domain and an oligonucleotide binding domain and it is thought that the ssDNA or ssRNA binds to only one subunit at any one time. Hydrolysis of NTP occurs in a step-wise fashion, in which one active site follows the other, driving the translocation of the helicase and the concomitant unwinding of the dsDNA or dsRNA. At present the exact mechanism by which hexameric helicases unravel duplex oligonucleotides remains uncertain.

An insight into why these helicases have evolved to encircle their substrate as multimers has been suggested based on their cellular functions.[48,51] The utilisation of hexameric helicases is strongly conserved across the three domains of life for tasks requiring high processivity, such as genome replication and homologous recombination. By corollary, monomeric helicases are used primarily for smaller-scale jobs. For example, the monomeric helicases UvrD[52] and PcrA[53] are involved in unwinding short stretches of dsDNA in DNA repair. This highlights the benefit to processivity of stringing the DNA or RNA through a protein ring so that it forms a tight link that does not readily dissociate after a catalytic event.

CONCLUSION

In conclusion—we suggest that dimeric and multimeric proteins that bind RNA and DNA have evolved as such for several reasons. In cases of oligonucleotide sequence recognition, multimers can increase oligonucleotide specificity and regulatory control. Where nonspecific interactions are required, protein multimerisation can facilitate the packaging or handling of long repeating units of oligonucleotide tracts. As further structural characterisation of protein-oligonucleotide complexes are added to our knowledge databases (particularly in the relatively uncharted realm of protein/RNA interactions), we can be confident that these themes will be continued and reiterated.

NOTE

All figures were generated using coordinates downloaded from the Protein Databank (www.rcsb.org) except for the dimer of dimers of RTP which was provided by the authors. The program PYMOL (DeLano, W.L. The PyMOL Molecular Graphics System (2002) on World Wide Web http://www.pymol.org) was used to generate the primary images which were then modified using GIMP (http://www.gimp.org/).

REFERENCES

1. Goodsell DS, Olson AJ. Structural symmetry and protein function. Annu Rev Biophys Biomol Struct 2000; 29:105-153.
2. Kurland CG. Translational accuracy and the fitness of bacteria. Annu Rev Genet 1992; 26:29-50.
3. Jones S, Daley DT, Luscombe NM et al. Protein-RNA interactions: a structural analysis. Nucleic Acids Res 2001; 29(4):943-954.
4. Suzuki M, Brenner SE, Gerstein M et al. DNA recognition code of transcription factors. Protein Eng 1995; 8(4):319-328.
5. Wolberger C. Multiprotein-DNA complexes in transcriptional regulation. Annu Rev Biophys Biomol Struct 1999; 28:29-56.
6. Auweter SD, Oberstrass FC, Allain FH. Sequence-specific binding of single-stranded RNA: is there a code for recognition? Nucleic Acids Res 2006; 34(17):4943-4959.
7. Katsamba PS, Park S, Laird-Offringa IA. Kinetic studies of RNA-protein interactions using surface plasmon resonance. Methods 2002; 26(2):95-104.
8. Bickle TA, Kruger DH. Biology of DNA restriction. Microbiol Rev 1993; 57(2):434-450.
9. Pingoud A, Fuxreiter M, Pingoud V et al. Type II restriction endonucleases: structure and mechanism. Cell Mol Life Sci 2005; 62(6):685-707.
10. Perona JJ, Martin AM. Conformational transitions and structural deformability of EcoRV endonuclease revealed by crystallographic analysis. J Mol Biol 1997; 273(1):207-225.
11. Kim YC, Grable JC, Love R et al. Refinement of Eco RI endonuclease crystal structure: a revised protein chain tracing. Science 1990; 249(4974):1307-1309.
12. Kaus-Drobek M, Czapinska H, Sokolowska M et al. Restriction endonuclease MvaI is a monomer that recognizes its target sequence asymmetrically. Nucleic Acids Res 2007; 35(6):2035-2046.
13. Khorasanizadeh S, Rastinejad F. Nuclear-receptor interactions on DNA-response elements. Trends Biochem Sci 2001; 26(6):384-390.
14. Wilson TE, Fahrner TJ, Milbrandt J. The orphan receptors NGFI-B and steroidogenic factor 1 establish monomer binding as a third paradigm of nuclear receptor-DNA interaction. Mol Cell Biol 1993; 13(9):5794-5804.
15. Luisi BF, Xu WX, Otwinowski Z et al. Crystallographic analysis of the interaction of the glucocorticoid receptor with DNA. Nature 1991; 352(6335):497-505.
16. Marmorstein R, Fitzgerald MX. Modulation of DNA-binding domains for sequence-specific DNA recognition. Gene 2003; 304:1-12.
17. Ptashne M. Gene regulation by proteins acting nearby and at a distance. Nature 1986; 322(6081):697-701.
18. Zhang RG, Joachimiak A, Lawson CL et al. The crystal structure of trp aporepressor at 1.8 A shows how binding tryptophan enhances DNA affinity. Nature 1987; 327(6123):591-597.
19. Lawson CL, Zhang RG, Schevitz RW et al. Flexibility of the DNA-binding domains of trp repressor. Proteins 1988; 3(1):18-31.
20. Otwinowski Z, Schevitz RW, Zhang RG et al. Crystal structure of trp repressor/operator complex at atomic resolution. Nature 1988; 335(6188):321-329.
21. Gajiwala KS, Burley SK. Winged helix proteins. Curr Opin Struct Biol 2000; 10(1):110-116.
22. Manna AC, Pai KS, Bussiere DE et al. The dimer-dimer interaction surface of the replication terminator protein of Bacillus subtilis and termination of DNA replication. Proc Natl Acad Sci USA 1996; 93(8):3253-3258.
23. Wilce JA, Vivian JP, Hastings AF et al. Structure of the RTP-DNA complex and the mechanism of polar replication fork arrest. Nat Struct Biol 2001; 8(3):206-210.

24. Smith MT, Wake RG. Definition and polarity of action of DNA replication terminators in Bacillus subtilis. J Mol Biol 1992; 227(3):648-657.
25. Neylon C, Kralicek AV, Hill TM et al. Replication termination in Escherichia coli: structure and antihelicase activity of the Tus-Ter complex. Microbiol Mol Biol Rev 2005; 69(3):501-526.
26. Mulcair MD, Schaeffer PM, Oakley AJ et al. A molecular mousetrap determines polarity of termination of DNA replication in E. coli. Cell 2006; 125(7):1309-1319.
27. Kornberg RD, Lorch Y. Twenty-five years of the nucleosome, fundamental particle of the eukaryote chromosome. Cell 1999; 98(3):285-294.
28. Ausio J. Histone variants—the structure behind the function. Brief Funct Genomic Proteomic 2006; 5(3):228-243.
29. Horn PJ, Peterson CL. Molecular biology. Chromatin higher order folding—wrapping up transcription. Science 2002; 297(5588):1824-1827.
30. Gollnick P. Regulation of the Bacillus subtilis trp operon by an RNA-binding protein. Mol Microbiol 1994; 11(6):991-997.
31. Antson AA, Dodson EJ, Dodson G et al. Structure of the trp RNA-binding attenuation protein, TRAP, bound to RNA. Nature 1999; 401(6750):235-242.
32. Neale MJ, Keeney S. Clarifying the mechanics of DNA strand exchange in meiotic recombination. Nature 2006; 442(7099):153-158.
33. Shin DS, Chahwan C, Huffman JL et al. Structure and function of the double-strand break repair machinery. DNA Repair (Amst) 2004; 3(8-9):863-873.
34. West SC. Molecular views of recombination proteins and their control. Nat Rev Mol Cell Biol 2003; 4(6):435-445.
35. Sun H, Treco D, Schultes NP et al. Double-strand breaks at an initiation site for meiotic gene conversion. Nature 1989; 338(6210):87-90.
36. Bell CE. Structure and mechanism of Escherichia coli RecA ATPase. Mol Microbiol 2005; 58(2):358-366.
37. Conway AB, Lynch TW, Zhang Y et al. Crystal structure of a Rad51 filament. Nat Struct Mol Biol 2004; 11(8):791-796.
38. Kinebuchi T, Kagawa W, Enomoto R et al. Structural basis for octameric ring formation and DNA interaction of the human homologous-pairing protein Dmc1. Mol Cell 2004; 14(3):363-374.
39. Shibata T, Nishinaka T, Mikawa T et al. Homologous genetic recombination as an intrinsic dynamic property of a DNA structure induced by RecA/Rad51-family proteins: a possible advantage of DNA over RNA as genomic material. Proc Natl Acad Sci USA 2001; 98(15):8425-8432.
40. Liu Y, West SC. Happy Hollidays: 40th anniversary of the Holliday junction. Nat Rev Mol Cell Biol 2004; 5(11):937-944.
41. Kelman Z, Finkelstein J, O'Donnell M. Protein structure. Why have six-fold symmetry? Curr Biol 1995; 5(11):1239-1242.
42. Yong J, Wan L, Dreyfuss G. Why do cells need an assembly machine for RNA-protein complexes? Trends Cell Biol 2004; 14(5):226-232.
43. Wilusz CJ, Wilusz J. Eukaryotic Lsm proteins: lessons from bacteria. Nat Struct Mol Biol 2005; 12(12):1031-1036.
44. Indiani C, O'Donnell M. The replication clamp-loading machine at work in the three domains of life. Nat Rev Mol Cell Biol 2006; 7(10):751-761.
45. Fay PJ, Johanson KO, McHenry CS et al. Size classes of products synthesized processively by DNA polymerase III and DNA polymerase III holoenzyme of Escherichia coli. J Biol Chem 1981; 256(2):976-983.
46. Oakley AJ, Prosselkov P, Wijffels G et al. Flexibility revealed by the 1.85 A crystal structure of the beta sliding-clamp subunit of Escherichia coli DNA polymerase III. Acta Crystallogr D Biol Crystallogr 2003; 59(Pt 7):1192-1199.
47. Gulbis JM, Kelman Z, Hurwitz J et al. Structure of the C-terminal region of p21(WAF1/CIP1) complexed with human PCNA. Cell 1996; 87(2):297-306.
48. Patel SS, Picha KM. Structure and function of hexameric helicases. Annu Rev Biochem 2000; 69:651-697.
49. Schaeffer PM, Headlam MJ, Dixon NE. Protein—protein interactions in the eubacterial replisome. IUBMB Life 2005; 57(1):5-12.
50. Putnam CD, Clancy SB, Tsuruta H et al. Structure and mechanism of the RuvB Holliday junction branch migration motor. J Mol Biol 2001; 311(2):297-310.
51. Hingorani MM, O'Donnell M. A tale of toroids in DNA metabolism. Nat Rev Mol Cell Biol 2000; 1(1):22-30.
52. Lee JY, Yang W. UvrD helicase unwinds DNA one base pair at a time by a two-part power stroke. Cell 2006; 127(7):1349-1360.
53. Soultanas P, Dillingham MS, Wiley P et al. Uncoupling DNA translocation and helicase activity in PcrA: direct evidence for an active mechanism. EMBO J 2000; 19(14):3799-3810.

CHAPTER 7

HOMO- AND HETERODIMERIZATION IN TRANSCRIPTIONAL REGULATION

Alister P. W. Funnell and Merlin Crossley*

School of Biotechnology and Biomolecular Sciences, University of New South Wales, Sydney, New South Wales, Australia; and, School of Molecular Bioscience, University of Sydney, Sydney, New South Wales, Australia

**Corresponding Author: Merlin Crossley—Email: m.crossley@unsw.edu.au*

Abstract: Eukaryotic transcription factors bind DNA and typically serve to localize large multiprotein complexes to particular genes to up- or downregulate transcription, thereby coordinating cellular responses to a variety of signals. Different combinations of transcription factors within DNA-binding multiprotein complexes allow individual proteins to partake in multiple different regulatory pathways. Many transcription factors can form homo- and heterodimers (or oligomers) with different partners, thus modulating DNA-binding specificity and affinity and/or the recruitment of different binding partners. This chapter reviews several of the mechanisms by which the homo- and heterodimerization of transcription factors contributes to transcriptional regulation.

INTRODUCTION

Intricate regulation of gene expression is crucial during the development of organisms. It dictates the multitude of varied cellular differentiation programs that higher eukaryotes exhibit. Control of gene expression can be asserted at numerous steps from the initial transcription of a gene through to the translation of a final protein product. However, despite these various layers of control, regulation of gene expression is primarily imposed at the level of transcription. That is to say, by factors which determine the frequency of transcriptional initiation.

Protein Dimerization and Oligomerization in Biology, edited by Jacqueline M. Matthews.

INDIVIDUAL GENES ARE CONTROLLED BY COMBINATIONS OF TRANSCRIPTION FACTORS

Transcription factors are DNA-binding, regulatory proteins that affect the rate of transcriptional initiation. They directly bind to control elements that are either situated immediately upstream of their target genes, known as promoters, or to more distal control elements such as enhancers. They then mediate their effects on transcriptional initiation through interactions with cofactors. These cofactors may modulate local chromatin configuration through the recruitment of: ATP-dependent chromatin remodeling factors such as SWI-SNF;[1] DNA-modifying enzymes such as DNA methyltransferases (DNMTs),[2] or; histone-modifying enzymes such as histone acetyltransferases (HATs), histone deacetylases (HDACs), histone methyltransferases (HMTs) and histone demethylases (Fig. 1).[3,4] Transcriptional cofactors may also operate by facilitating or inhibiting the assembly of the transcription pre-initiation complex (PIC). For genes transcribed by RNA polymerase II, the PIC comprises the so called 'general transcription factors:' TFIIA, TFIIB, TFIID, TFIIE, TFIIF, TFIIH and RNA polymerase II (Fig. 1). The PIC constitutes the transcriptional apparatus that is required for a basal level of gene expression. In eukaryotes, changes in chromatin structure and in the rate of PIC assembly are largely brought about by the complex interplay of multiple transcription factors and cofactors.

The sequencing of the human genome brought with it the somewhat humbling estimation that there are roughly only 25,000 human genes.[5] In comparison, mice have a similar number of genes[6] and the genome of the nematode *Caenorhabditis elegans* contains approximately 19,000 genes.[7] Thus, there is often a poor correlation between the gene number and physical and behavioral complexity of an organism. It has since been postulated that both the relative and absolute number of transcription factors in an organism's genome contribute to its complexity.[8] In eukaryotes, transcription factors interact and function in a combinatorial manner to regulate gene expression. This means that a modest increase in the number of transcription factors in an organism can lead to a large number of novel combinations of

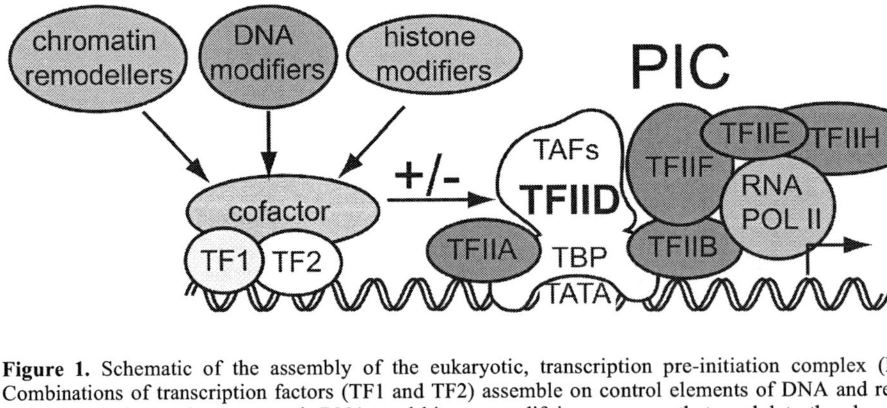

Figure 1. Schematic of the assembly of the eukaryotic, transcription pre-initiation complex (PIC). Combinations of transcription factors (TF1 and TF2) assemble on control elements of DNA and recruit cofactors. Cofactors in turn recruit DNA- and histone-modifying enzymes that modulate the chromatin state of the locus and accessibility for PIC formation. Cofactors may also directly promote or impede the recruitment of PIC components. The PIC consists of the general transcription factors TFIIA, TFIIB, TFIID, TFIIE, TFIIF, TFIIH and RNA polymerase II. TFIID comprises TATA-binding protein (TBP), through which it binds DNA, and an assortment of TBP-associated factors (TAFs).

transcription factors that could form new regulatory pathways. This type of combinatorial control allows the development of complex and intricate regulatory networks. In contrast, prokaryotes often utilize single transcription factors or simple homodimers to regulate the expression of genes.[9] See also Chapter 6 by Wilce et al.

Eukaryotic RNA polymerase II transcription typically occurs in the presence of multiple factors that individually act in a positive or negative manner.[10] These multiprotein complexes are sometimes termed enhanceosomes[11] and enable exquisite control of gene expression. Many transcription factors have been shown to be capable of oligomerizing (most commonly forming homodimers) and to date, numerous crystal structures have been obtained for multimeric transcription factor complexes bound to DNA (for reviews see 12-14). In this chapter we will look at how transcription factors employ both homo and hetero-oligomerization to both positively and negatively regulate expression of specific target genes in order to define discrete regulatory pathways.

ACHIEVING DNA-BINDING THROUGH TRANSCRIPTION FACTOR DIMERIZATION

In eukaryotes, there are only a score or so of classes of DNA-binding domains, including the zinc finger, the basic region-leucine zipper (bZIP), the helix-turn-helix and homeodomain,[15] the ETS domain,[16] the MADS box,[17] the basic helix-loop-helix (bHLH),[18] the Rel homology domain[19] and the immunoglobulin-like fold.[20,21]

Several of these domains allow transcription factors to bind DNA as monomers. This is often the case for zinc finger-, ETS domain- and homeodomain-containing transcription factors. It should be noted, however, that the presence of these domains does not preclude dimerization and many transcription factors that contain these domains do form dimers or higher order oligomers. In some instances, the tandem repetition of the domain serves a similar purpose to dimerization by increasing the available contacts the protein makes with the DNA to facilitate an increase in both the specificity and affinity of binding. This is often the case for classical zinc fingers, where three closely spaced fingers are usually required for high affinity binding to DNA.

In contrast, other domains, such as the bZIP, bHLH, MADS box and the Rel homology domain, require noncovalent dimerization in order to bind DNA. Dimerization of transcription factors can occur at surfaces that are distant from the DNA-binding domains. For example, signal transducer and activator of transcription proteins (STATs) dimerize by interaction of phosphorylated SH2 domains. The crystal structure of a STAT1 homodimer bound to DNA reveals that the C-terminal SH2 dimerization interface is spatially isolated from the central, immunoglobulin-like DNA-binding domain (Fig. 2A).[22]

For the most part, however, if dimerization is required for DNA-binding, then it occurs through the DNA-binding or adjacent domains. Dimerization of these domains stabilizes the structure and induces conformational changes that enable the required contacts to be made with the major groove of DNA. For example, bHLH transcription factors, such as MyoD, dimerize by means of their helix-loop-helix domains.[18,23] Following dimerization, a basic region adjacent to the HLH makes contacts with DNA, but does not cooperate in and is not required for dimerization (Fig. 2B).[23]

The basic leucine zipper (bZIP) motif binds DNA in a similar fashion (Fig. 2C). The leucine zipper consists of a pair of long α-helices that contain a seven-amino acid motif

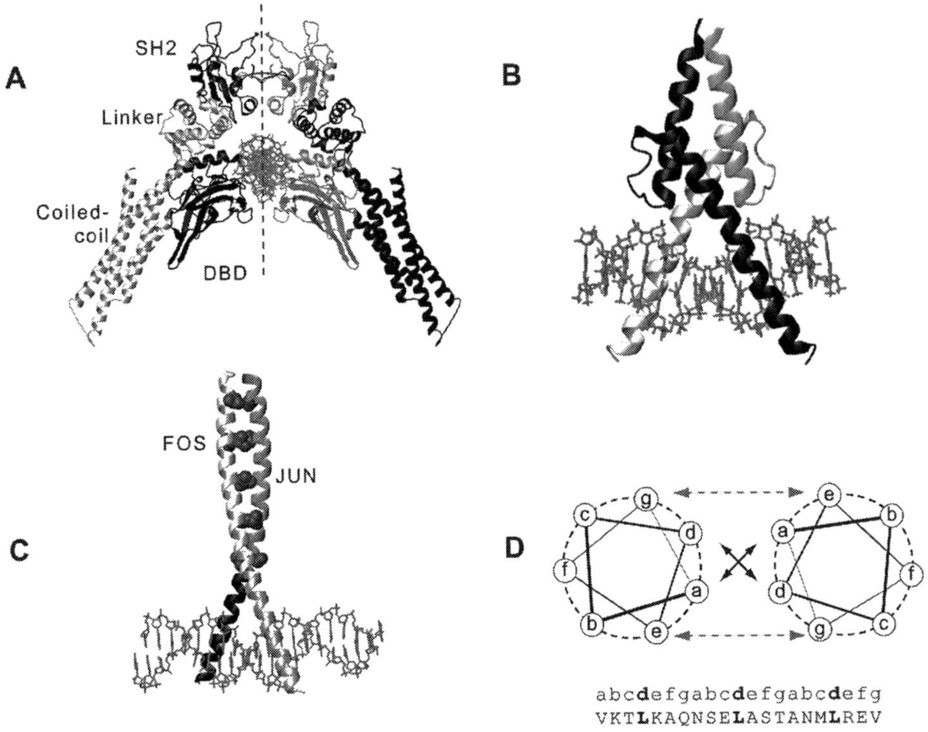

Figure 2. Structures of homo and heterodimeric transcription factors. A) Structure of phosphorylated STAT1 bound to DNA (PDB accession code 1BF5) as a homodimer. The coiled-coil, DNA-binding (DBD), linker and SH2 domains are indicated. B) Structure of MyoD bHLH homodimer bound to DNA (PDB accession code 1MDY). The two subunits are shown in different colours. C) Structure of a FOS/JUN bZIP heterodimer bound to DNA (PDB accession code 1MDY). The protein chains are indicated and the basic domains in darker colors bind DNA. The sidechains of the leucine residues are shown in CPK format. D) Schematic of the heptad repeat that forms the basis of the leucine zipper. Part of the leucine zipper sequence for JUN is shown below the heptad repeat letter designations. A color version of this image is available at www.landesbioscience.com/curie.

known as a heptad-repeat (a, b, c, d, e, f and g; Fig. 2D).[24] There are commonly five or six heptads in human leucine zippers, although bZIP transcription factors containing as few as three and as many as nine have been identified.[25] Position **d** is typically the leucine by which the motif gets its name, while position **a** is normally another hydrophobic residue.[26] Dimerization is accomplished by the 'zippering' of the **a** and **d** position residues to form the hydrophobic core of a parallel coiled-coil (Fig. 2C).[27] Upon dimerization of two bZIP proteins, the basic regions, which are N-terminal and adjacent to the leucine zipper, are brought into close proximity, enabling binding to DNA.[28,29]

There are more than 50 human genes that encode proteins with bZIP motifs.[25,30] In theory, this could produce ~1500 different possible homo- and heterodimers. However, in biology, bZIP proteins show preferences for specific dimerization partners based on the amino acid compositions of their leucine zippers. To explore this, predictive algorithms have been devised based on experimental observations of bZIP partner preferences.[25,31-36] For instance, at position **a** of a heptad, an asparagine residue is more likely to promote homodimerization than a lysine residue, which promotes heterodimerization.[37] In addition, charged amino acids at the **e** and **g** positions of one bZIP protein may engage in attractive or repulsive interactions with charged amino acids at the **g** and **e** positions respectively of another bZIP protein (Fig. 2D).[38] For example, a bZIP protein that has acidic residues at its **e** positions is not likely to dimerize with a bZIP protein with acidic residues at its **g** positions. Rather, dimerization would be preferred with a bZIP protein with basic residues at its **g** positions due to favourable electrostatic interactions. This is illustrated by the dimerization preferences of the proto-oncoproteins FOS and JUN. FOS/JUN heterodimers form readily and demonstrate strong DNA-binding activity.[39] JUN can also form homodimers that bind DNA. However, FOS homodimers are unstable and do not exhibit detectable DNA-binding activity.[39] The explanation for these dimerization preferences lies in the charged residues at the **e** and **g** positions of FOS and JUN. A JUN homodimer has two unfavorable **g-e** charge repulsions, whereas a FOS homodimer has four.[40]

A recent study has shown that, based on available experimental data, this predictive methodology is accurate in over 90% of cases.[35] Another recent study has collated predictions and experimental results to create an extensive map of interactions between human bZIP proteins.[36] The authors predict that ~340 of the possible 1500 or so bZIP dimer combinations are likely to occur under physiological conditions.

From the many interactions within the bZIP superfamily alone, one can appreciate that there are a vast number of transcription factor oligomers that may define the complex regulatory pathways of higher eukaryotes. It should also be noted that many transcription factors can interact with transcription factors from other superfamilies. For example, the FOS/JUN heterodimer further interacts with nuclear factor of activated T-cells (NFAT) to form a higher order, trimeric DNA-bound complex.[41] NFAT binds to DNA by means of two immunoglobulin-like folds.[41] Moreover, the MADS box serum response factor (SRF) has been demonstrated to interact with transcription factors from a dozen or so superfamilies.[42]

There are many ramifications from the homo- and hetero-oligomerization of transcription factors, including changes in DNA-binding affinity, alterations of sequence specificity and variations in modes of transcriptional regulation.

ALTERING DNA-BINDING AFFINITY THROUGH SELECTIVE DIMERIZATION

One common consequence of transcription factor oligomerization is an increase in DNA-binding affinity. Whereas some transcription factor superfamilies require dimerization for efficient DNA-binding, others are capable of binding DNA as monomers, but interact with other transcription factors that modulate DNA-binding properties. For example, the yeast homeodomain protein MATα2 can bind to DNA as a monomer.[43] However, MATα2 can heterodimerize with another homeodomain protein, MATa1,[44] or form a heterotetramer with the MADS box protein MCM1.[45,46] The MATα2/MATa1 heterodimer and the MATα2/MCM1 heterotetramer bind to DNA with much stronger affinity and sequence specificity than the MATα2 monomer alone (Fig. 3A).[47] Such a phenomenon is also observed for the FOS/JUN/NFAT trimer.[12,41,48] Indeed, significant transcriptional activation of some NFAT target genes is dependent on this trimeric complex forming at their promoters.[49] Cooperativity of DNA-binding is also observed among the general transcription factors. TFIIB is now thought to bind to TFIIB recognition elements (BREs) flanking the TATA box (Fig. 3B).[50] This increases the affinity with

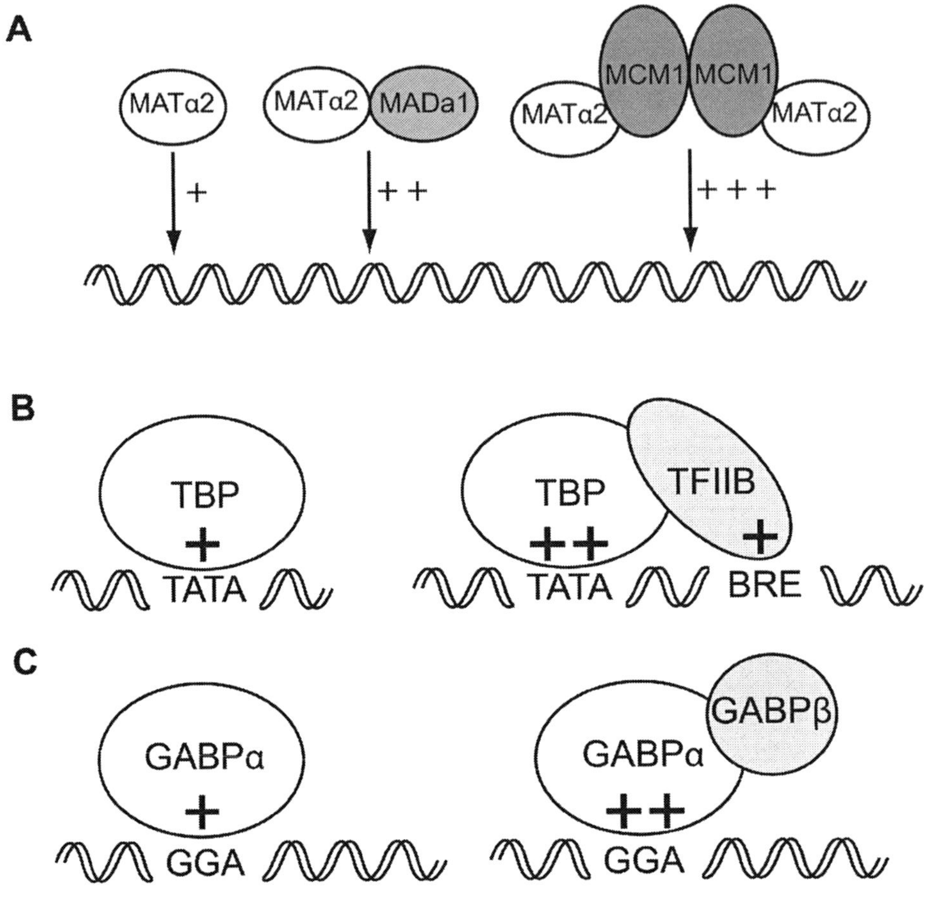

Figure 3. Dimerization can increase DNA-binding affinity. A) MATα2 oligomers (MATα2/MADa1 heterodimer and MATα2/MCM1 heterotetramer) bind more avidly to DNA than MATα2 monomers. B) General transcription factor TFIIB binds to TFIIB recognition elements (BREs) adjacent to TATA boxes and increases the affinity with which TATA-binding protein (TBP) binds DNA. C) The nonDNA-binding protein GABPβ increases the strength of the interaction between GABPα and GGA sites of DNA.

which the TATA-binding protein (TBP) subunit of TFIID binds DNA and thus promotes PIC assembly.[51]

In some circumstances non DNA-binding proteins can also augment the DNA-binding affinity of transcription factors. GA-binding protein α (GABPα) is important in the regulation of nuclear mitochondrial genes and recognizes GGA motifs by means of an ETS domain[52,53] GABPα interacts with a structurally unrelated factor, GABPβ, which lacks a DNA-binding domain. Residues found in and C-terminal to, the ETS domain of GABPα interact with an ankyrin repeat domain of GABPβ to mediate this dimerization.[52] The resulting GABPα/β dimer binds DNA with higher affinity than GABPα alone (Fig. 3C).[52] The explanation for this increased DNA-binding affinity is not well understood, however, it has been observed that Lys69 of GABPβ forms a hydrogen bond with Gln321 of GABPα and this may stabilize the contact made between Gln321 and a DNA phosphate group.[54] A hydrogen bond is also believed to be formed and stabilized between Leu322 of GABPα and a DNA minor groove phosphate group upon dimerization with GABPβ.[14]

In a similar manner, the DNA-binding affinity of RUNX1 (Runt-related protein 1) via the Runt domain is enhanced by binding CBFβ (core-binding factor β), which lacks a DNA-binding motif.[55] Heterodimerization induces conformational changes in RUNX1 that increase the extent of hydrogen-bonding made between that protein and the major and minor grooves of DNA.[14,55,56]

MODULATING DNA SEQUENCE SPECIFICITY THROUGH DIMERIZATION

Perhaps the most significant implication of transcription factor oligomerization is the modulation of DNA sequence specificity. Combinations of DNA-binding transcription factors can recognize extended and hence more specific, DNA elements in the control regions of genes. In addition, oligomerization of unrelated transcription factors can result in complexes that recognize diverse, novel DNA elements. These novel elements can be distinct from the DNA sequences recognized by the monomers that constitute the complex. Because many transcription factors exhibit spatio- and temporal-restricted expression patterns, different transcription factor combinations form in different cell types and at different developmental stages. Oligomerization thus allows the formation of intricate regulatory networks that are essential for the fine-tuning of gene expression programs in higher eukaryotes.

The simplest case of increased specificity is manifest in oligomers that recognize DNA elements that are essentially the juxtaposition of the recognition elements of the individual monomers. The NF-κB dimers represent such an example. The NF-κB p50 and p65 proteins both contain a Rel homology domain which allows binding to palindromic DNA sites upon homodimerization.[57] In a p50 homodimer, each p50 subunit recognizes a GGGRN half-site of DNA, where R is a purine.[58-60] In contrast, in a p65 homodimer, each p65 subunit of a homodimer recognizes a GGAA motif.[60,61] In both instances, these half sites are separated by a single base pair. Hence a p50 homodimer recognizes the sequence GGGRN A/T NYCCC, where Y is a pyrimidine and a p65 homodimer recognizes GGAA N TTCC (Fig. 4A).[60,61] A p50/p65 heterodimer binds to a composite sequence that retains the half-site recognition elements of the two monomers—a GGGRN N TTCC site (Fig. 4A).[60] Thus, each of the three dimers has the capacity to regulate a distinct subset of genes.

For other transcription factors, the presence of alternative binding partners can create novel recognition sites that do not (or only weakly) resemble the recognition sites of the individual monomers. STAT homo- and heterodimers bind to the palindromic consensus site TTCNNNGAA, known as the interferon gamma activated site (GAS) (Fig. 4B).[62] In a STAT1/STAT2 heterodimer, both STAT1 and STAT2 contribute to DNA-binding. This heterodimer can bind interferon regulatory factor 9 (IRF9) to form a higher order complex, which is known as interferon-stimulated gene factor 3 (ISGF3).[21,63] This complex binds to interferon-stimulated response elements (ISREs) of the consensus AGTTTNNNTTTC (Fig. 4B). Within ISGF3, STAT1 still actively binds DNA in conjunction with IRF9. In contrast, STAT2 is not engaged in DNA-binding but acts as a potent transcriptional activation domain. Thus, the DNA element that is bound is considerably different to that bound by a STAT1/STAT2 heterodimer, which allows different complexes containing both STAT1 and STAT2 to regulate the expression of distinct subsets of genes.

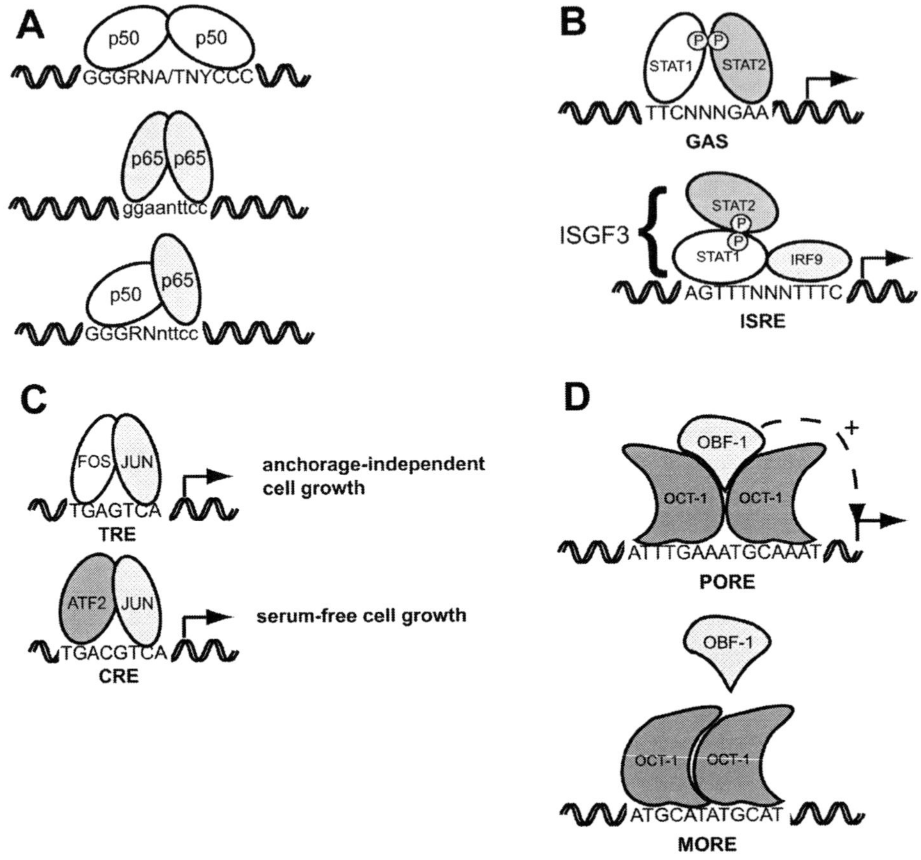

Figure 4. Dimerization can alter DNA-binding sequence specificity. A) The NF-κB p50 homodimer binds to two GGGRN half-sites of DNA while the p65 homodimer binds to two GGAA half-sites of DNA. A p50/p65 heterodimer binds to an element (GGGRN N TTCC) that contains both of these half-sites. B) Phosphorylated STAT1 and STAT2 heterodimerize and bind the interferon gamma activated site (GAS) DNA element, TTCNNNGAA. The STAT1/STAT2 heterodimer can also interact with interferon regulatory factor 9 (IRF9) to form the interferon-stimulated gene factor 3 (ISGF3) complex. ISGF3 binds to interferon-stimulated response elements (ISREs) of the consensus AGTTTNNNTTTC. C) The FOS/JUN heterodimer binds to the TPA-responsive element (TRE) of the consensus TGAGTCA, while the JUN/ATF2 heterodimer recognizes the cAMP-responsive element (CRE) of the sequence TGACGTCA. D) OCT-1 homodimers bind to the palindromic OCT factor recognition element (PORE; ATTTGAAATGCAAAT), or the MORE motif (ATGCATATGCAT), depending on the dimerization face employed. An OCT-1 homodimer bound to a PORE motif can recruit the transcriptional coactivator OBF-1, while an OCT-1 homodimer bound to a MORE motif cannot.

Modulation of DNA-binding sequence specificity is also achieved by alternative dimerization of members within the bZIP superfamily. The FOS/JUN heterodimer preferentially binds to TPA-responsive elements (TREs) of the consensus TGAGTCA (Fig. 4C).[36,64] JUN also readily dimerizes with another bZIP protein, ATF2. The JUN/ATF2 heterodimer does not recognize a TRE sequence, but rather, a cAMP-responsive element (CRE) of the consensus TGACGTCA (Fig. 4C).[36,65] In addition to binding different DNA elements, the two heterodimers have distinct biological roles. The FOS/JUN dimer promotes anchorage independent cell growth, whereas the JUN/ATF2 dimer promotes serum-free growth in vitro.[66] Another interesting family of transcription factors that use dimerization to expand their repertoire of DNA recognition elements is the nuclear receptor family, which is discussed in more detail in Chapter 6 by Wilce et al.

Some transcription factor dimers are capable of binding to different DNA elements depending on the various possible protein interfaces that they employ to dimerize. The ubiquitous OCT-1 transcription factor is a POU domain protein that can homo- and heterodimerize and is implicated in the regulation of immunoglobulin heavy chain genes.[67] OCT-1 can form homodimers through different surfaces.[68] Consequently, one OCT-1 homodimer configuration recognizes a palindromic OCT factor recognition element (PORE; ATTTGAAATGCAAAT; Fig. 4D),[68,69] whereas another configuration of the OCT-1 homodimer binds a different motif (MORE; ATGCATATGCAT; Fig. 4D).[68] An OCT-1 homodimer bound to a PORE motif is able to recruit the B cell specific transcriptional coactivator OBF-1 (Fig. 4D), but cannot do so when bound to a MORE motif.[68] This is because the same surface of OCT-1 is used to form a MORE-binding homodimer and to bind OBF1. Thus, different transcription factor oligomers enable varied modes of transcriptional regulation.

CONVERTING ACTIVATORS INTO REPRESSORS BY CHANGING PARTNERS

Eukaryotic transcription factors are modular in structure and in addition to DNA-binding domains, possess activation and/or repression domains that affect the rate of transcription, usually through recruitment of cofactors. Transcription factors may therefore interact with other transcription factors that can up- or downregulate transcription. In other words, dimerization with different partner proteins can permit contrasting effects on transcriptional control. A well-characterized model of this type of combinatorial control is the MYC/MAX/MAD network.

The MYC, MAX and MAD transcription factors all belong to the basic helix-loop-helix leucine zipper (bHLHZ) family. Like the bZIP proteins, dimerization is obligatory if these transcription factors are to bind DNA. Members of the bHLHZ family also show distinct binding partner preferences. For example, MYC is incapable of forming homodimers under physiological conditions,[70] but readily heterodimerizes with MAX.[71] The MYC/MAX dimer binds strongly to the DNA E-box motif, CACGTG (Fig. 5A).[71,72] MYC contains a potent transactivation domain that recruits numerous coactivators such as TRRAP, an element of a histone acetyl transferase complex.[73] The MYC/MAX heterodimer therefore activates transcription of target genes (Fig. 5A).[74,75]

MAX can also interact with the transcriptional repressor MAD. The MAD/MAX dimer also binds to E-boxes. However, in contrast to MYC, MAD contains a potent transcriptional repression domain known as a SIN3 interaction domain (SID).[76] Through this domain, MAD recruits the corepessor SIN3, a component of a large histone deacetylase complex (Fig. 5B). The MYC/MAX and MAD/MAX dimers therefore compete for target sites and antagonize each other's function.[77,78]

Many target genes of the MYC/MAX and MAD/MAX complexes have been proposed and a high proportion of these are involved in cell cycle regulation.[79,80] Due

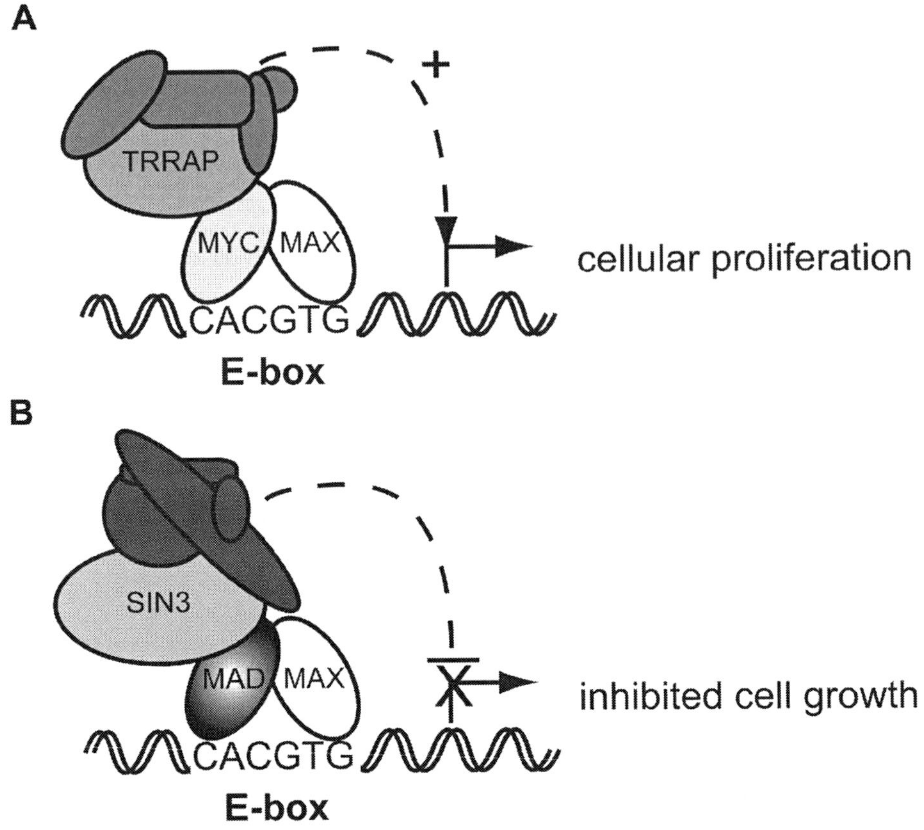

Figure 5. Alternative heterodimerization of MAX alters its effect on transcription. A) The MYC/MAX heterodimer binds to the E-box, CACGTG and activates transcription through the recruitment of coactivators such as TRRAP. B) The MAD/MAX heterodimer also binds to the E-box but represses transcription through the recruitment of histone deacetylases such as SIN3.

to their opposing molecular roles, MYC stimulates cellular proliferation while MAD expression is associated with terminal differentiation and blocked cell growth. This is illustrated during the differentiation of HL-60 myoblast cells, where MYC/MAX dimers are replaced by MAD/MAX dimers at the *cyclin D2* promoter.[81] This is accompanied by an increase in HDAC1 recruitment and a concomitant loss of histone acetylation and RNA pol II recruitment.[81] Such interchange of functionally opposing binding partners is widely used to modulate gene expression.

DOMINANT NEGATIVES: DECOYS THAT PREVENT THE FORMATION OF FUNCTIONAL DIMERS

Dominant negative proteins bind to and abrogate the function of their cognate binding partners and thus can be seen as molecular switches that can rapidly impact on gene regulatory networks. In many cases, dominant negative proteins bind transcription factors and inhibit DNA-binding activity, either by occluding or disrupting the DNA-binding site, or by preventing dimerization with other transcription factors that are required for DNA-binding. Several such dominant negative proteins have been characterized to date.

The CCAAT/enhancer binding proteins (C/EBPs) are transcription factors that have roles in adipogenesis, hematopoiesis and cell cycle regulation. They belong to the bZIP superfamily and must homo- or heterodimerize in order to bind DNA. Their DNA-binding activity, however, has been found to be attenuated by a related bZIP protein, C/EBP homologous protein 10 (CHOP).[82] CHOP contains a bZIP motif that enables dimerization with C/EBPs. However, in contrast to C/EBPs, CHOP contains proline and glycine residues in the basic region adjacent to the leucine zipper.[82,83] These residues abrogate CHOP's ability to bind DNA (Fig. 6).[82]

CHOP is a dynamic regulator of the C/EBP family during adipogenesis.[82,84] It is induced early in adipogenesis and sequesters C/EBPβ until mitotic clonal expansion of pre-adipocytes occurs. CHOP is subsequently downregulated and C/EBPβ is liberated.[85,86] Thereupon, C/EBPβ is able to bind to and activate target genes such as *C/EBPα*. This process needs to be tightly coordinated, because C/EBPα is antimitotic and its premature expression would inhibit adipogenesis.

Another well-characterized example is the inhibitor of DNA-binding protein, ID. ID is a HLH protein that interacts with other bHLH transcription factors. However, ID lacks the basic region adjacent to the HLH motif and hence heterodimers containing ID are incapable of binding DNA.[87] ID expression prevents differentiation and maintains the pluripotency of stem cells.[88] Lineage commitment and differentiation of progenitor cells is associated with downregulation of ID and overexpression of ID has been associated with many types of cancer.[89-91]

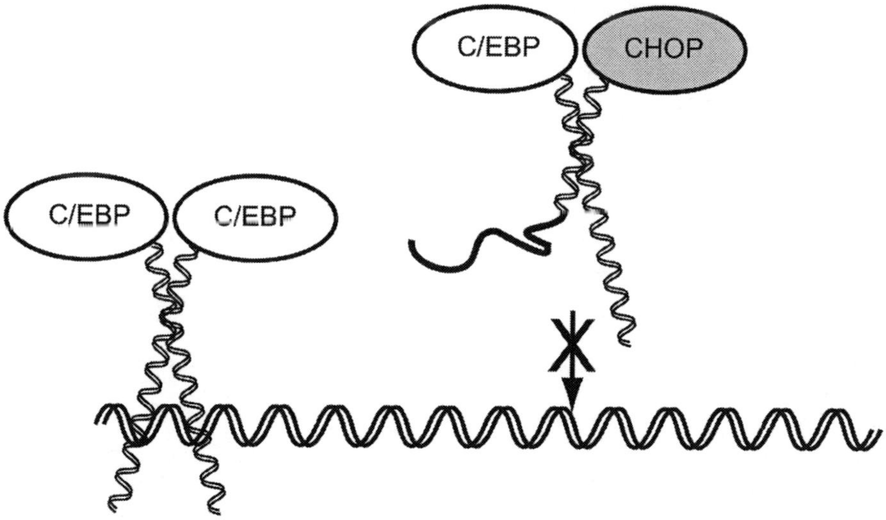

Figure 6. Dominant negative proteins can inhibit the activity of transcription factors by impeding DNA-binding. CCAAT/enhancer binding protein (C/EBP) dimers are capable of binding DNA while C/EBP/CHOP heterodimers are unable to bind DNA.

OLIGOMERIZATION WITHIN THE GENERAL TRANSCRIPTION FACTORS

Several mechanisms are employed by the general transcription factors to prevent widespread nonspecific assembly of the PIC. Numerous observations have shown that TBP forms homodimers that occlude the DNA-binding domain to inhibit TATA-binding (Fig. 7A).[51] Similarly, TAF1 and BTAF1 can both bind to the concave DNA-binding region of TBP and thus block TATA-binding (Fig. 7A).[92,93] Furthermore, TAF1 can also bind to a separate region of TBP which is normally engaged in TBP-TFIIA interactions.[94] Such binding prevents PIC assembly by blocking the recruitment of the general transcription factor TFIIA.

In addition, some general transcription factors can exert their negative effects on PIC assembly after TBP has bound to DNA. For example, BTAF1 can interact with DNA-bound TBP through an exposed region of TBP.[93] BTAF1 possesses enzymatic activity and causes the dissociation of TBP from DNA in an ATP-dependent fashion (Fig. 7B).[95] A further example is negative cofactor 2 (NC2), which binds to the underside of the TBP-TATA complex and acts as a molecular clamp (Fig. 7C). The crystal structure of the NC2-TBP-TATA ternary complex reveals that NC2 sterically blocks the subsequent binding of TFIIA and TFIIB required for PIC assembly.[13,96] These diverse regulatory mechanisms are thought to be employed to reduce leaky PIC assembly and hence transcription, from nonpromoter regions.

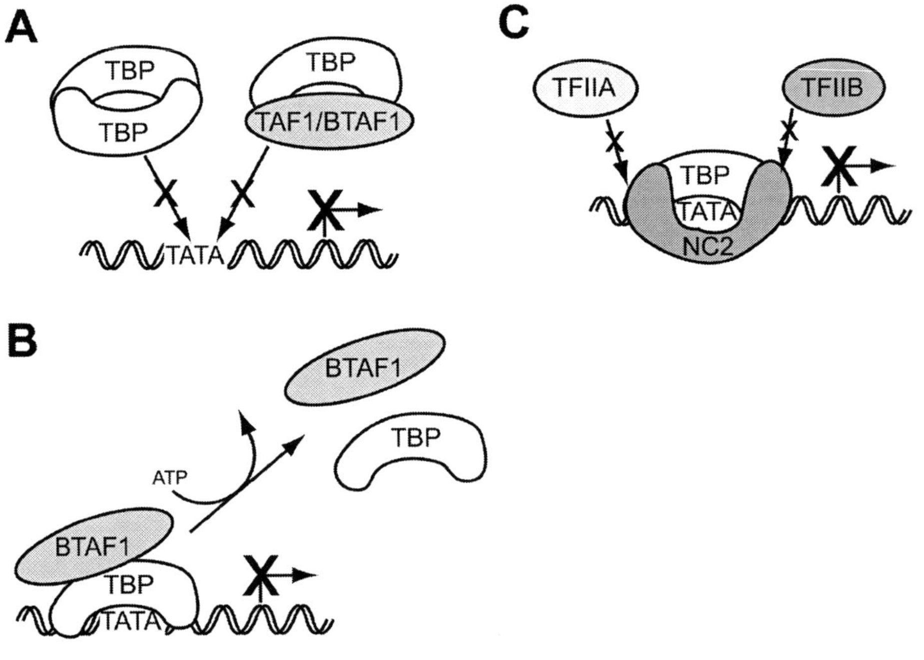

Figure 7. The general transcription factors oligomerize to prevent nonspecific PIC formation. A) The TBP homodimer and the TBP/TAF1 and TBP/BTAF1 heterodimers are unable to bind the TATA-box of DNA. B) BTAF1 interacts with DNA-bound TBP and catalyzes the ATP-dependent dissociation of TBP from DNA. C NC2 binds to DNA-bound TBP and prevents further interactions with TFIIA and TFIIB.

THE REGULATION OF DIMERIZATION

It is emerging that there are a number of factors that regulate transcription factor dimerization. For instance, posttranslational modifications can enhance or inhibit dimerization. STAT transcription factors, for example, require phosphorylation for dimerization, nuclear translocation and transcriptional activity. The binding of ligands can also be an important determinant of dimerization, as exemplified by the nuclear receptor transcription factors discussed in Chapter 6 by Wilce et al. Lastly, the dimerization of some transcription factors is promoted by auxiliary proteins. Such an auxiliary protein is bZIP enhancing factor (BEF), which exhibits chaperone-like properties and enhances the folding of bZIP monomers to form dimers.[97]

HIGHER ORDER MULTIMERIZATION

Transcription factors usually regulate gene expression as part of large multiprotein complexes. The recruitment and formation of such higher order complexes is often preceded and mediated by the binding of monomeric or dimeric transcription factors to control elements of DNA. There are, however, examples of transcription factor homo- and heteromultimers that form prior to DNA-binding and even require multimerization for DNA-binding.

NF-Y is a heterotrimeric transcription factor complex made up of three distinct subunits, NF-YA, NF-YB and NF-YC (reviewed in ref. 98). The NF-Y subunits are each incapable of binding DNA as monomers, but contain at least two protein-interaction domains. NF-YB and NF-YC interact through a histone fold motif, which then enables NF-YA to bind by means of a subunit interaction domain to form a heterotrimer. The trimeric form binds with high affinity to the widespread CCAAT element in control regions of DNA. Once bound to DNA, the trimer recruits cofactors through glutamine-rich domains[99] and interacts with the general transcriptional apparatus via TBP-binding domains.[100]

The heat shock transcription factors (HSFs) constitute another example of this phenomenon (reviewed in refs. 101-103). Eukaryotes respond to thermal and other environmental stresses with the production of heat shock proteins (HSPs). Gene expression of *HSP*s is regulated by HSFs. In humans, intramolecular interactions between coiled-coil domains of the HSF1 monomer render the protein transcriptionally inactive because it is unable to bind DNA. However, in the presence of external stimuli such as heat stress, HSF1 rapidly forms homotrimers that are able to bind heat shock elements (HSEs) in the control regions of *HSP* genes and subsequently activate transcription.

CONCLUSION

This chapter has focused on how homo- and hetero-oligomerization is employed by transcription factors to modulate DNA-binding affinity and specificity and to dynamically regulate the activation or repression of target genes. Oligomerization of eukaryotic transcription factors enables combinatorial control of gene regulation. This versatile mechanism enables a plethora of diverse gene regulatory pathways that are required to orchestrate the many differentiation programs present in metazoans. Predictive methodology and experimental data continue to define the myriad transcription factor multimers with biological roles and to identify the molecular bases of higher order complexity.

REFERENCES

1. Mellor J. Dynamic nucleosomes and gene transcription. Trends Genet 2006; 22(6):320-329.
2. Siedlecki P, Zielenkiewicz P. Mammalian DNA methyltransferases. Acta Biochim Pol 2006; 53(2):245-256.
3. Verdone L, Agricola E, Caserta M et al. Histone acetylation in gene regulation. Brief Funct Genomic Proteomic 2006; 5(3):209-221.
4. Shi Y, Whetstine JR. Dynamic regulation of histone lysine methylation by demethylases. Mol Cell 2007; 25(1):1-14.
5. Finishing the euchromatic sequence of the human genome. Nature 2004; 431(7011):931-945.
6. Waterston RH, Lindblad-Toh K, Birney E et al. Initial sequencing and comparative analysis of the mouse genome. Nature 2002; 420(6915):520-562.
7. Genome sequence of the nematode C. elegans: a platform for investigating biology. Science 1998; 282(5396):2012-2018.
8. Levine M, Tjian R. Transcription regulation and animal diversity. Nature 2003; 424(6945):147-151.
9. Remenyi A, Scholer HR, Wilmanns M. Combinatorial control of gene expression. Nat Struct Mol Biol 2004; 11(9):812-815.
10. Roeder RG. Transcriptional regulation and the role of diverse coactivators in animal cells. FEBS Lett 2005; 579(4):909-915.
11. Kim TK, Maniatis T. The mechanism of transcriptional synergy of an in vitro assembled interferon-beta enhanceosome. Mol Cell 1997; 1(1):119-129.
12. Wolberger C. Multiprotein-DNA complexes in transcriptional regulation. Annu Rev Biophys Biomol Struct 1999; 28:29-56.
13. Burley SK, Kamada K. Transcription factor complexes. Curr Opin Struct Biol 2002; 12(2):225-230.
14. Ogata K, Sato K, Tahirov TH. Eukaryotic transcriptional regulatory complexes: cooperativity from near and afar. Curr Opin Struct Biol 2003; 13(1):40-48.
15. Harrison SC. A structural taxonomy of DNA-binding domains. Nature 1991; 353(6346):715-719.
16. Wasylyk B, Hahn SL, Giovane A. The Ets family of transcription factors. Eur J Biochem 1993; 211(1-2):7-18.
17. Shore P, Sharrocks AD. The MADS-box family of transcription factors. Eur J Biochem 1995; 229(1):1-13.
18. Garrell J, Campuzano S. The helix-loop-helix domain: A common motif for bristles, muscles and sex. Bioessays 1991; 13(10):493-498.
19. Rushlow C, Warrior R. The rel family of proteins. Bioessays 1992; 14(2):89-95.
20. Horvath CM, Wen Z, Darnell JE Jr. A STAT protein domain that determines DNA sequence recognition suggests a novel DNA-binding domain. Genes Dev 1995; 9(8):984-994.
21. Lim CP, Cao X. Structure, function and regulation of STAT proteins. Mol Biosyst 2006; 2(11):536-550.
22. Chen X, Vinkemeier U, Zhao Y et al. Crystal structure of a tyrosine phosphorylated STAT-1 dimer bound to DNA. Cell 1998; 93(5):827-839.
23. Davis RL, Cheng PF, Lassar AB et al. The MyoD DNA binding domain contains a recognition code for muscle-specific gene activation. Cell 1990; 60(5):733-746.
24. McLachlan AD, Stewart M. Tropomyosin coiled-coil interactions: Evidence for an unstaggered structure. J Mol Biol 1975; 98(2):293-304.
25. Vinson C, Acharya A, Taparowsky EJ. Deciphering B-ZIP transcription factor interactions in vitro and in vivo. Biochim Biophys Acta 2006; 1759(1-2):4-12.
26. Landschulz WH, Johnson PF, McKnight SL. The leucine zipper: A hypothetical structure common to a new class of DNA binding proteins. Science 1988; 240(4860):1759-1764.
27. O'Shea EK, Klemm JD, Kim PS et al. X-ray structure of the GCN4 leucine zipper, a two-stranded, parallel coiled coil. Science 1991; 254(5031):539-544.
28. Patel L, Abate C, Curran T. Altered protein conformation on DNA binding by Fos and Jun. Nature 1990; 347(6293):572-575.
29. Weiss MA, Ellenberger T, Wobbe CR et al. Folding transition in the DNA-binding domain of GCN4 on specific binding to DNA. Nature 1990; 347(6293):575-578.
30. Tupler R, Perini G, Green MR. Expressing the human genome. Nature 2001; 409(6822):832-833.
31. Fassler J, Landsman D, Acharya A et al. B-ZIP proteins encoded by the Drosophila genome: evaluation of potential dimerization partners. Genome Res 2002; 12(8):1190-1200.
32. Vinson C, Myakishev M, Acharya A et al. Classification of human B-ZIP proteins based on dimerization properties. Mol Cell Biol 2002; 22(18):6321-6335.
33. Newman JR, Keating AE. Comprehensive identification of human bZIP interactions with coiled-coil arrays. Science 2003; 300(5628):2097-2101.
34. Deppmann CD, Acharya A, Rishi V et al. Dimerization specificity of all 67 B-ZIP motifs in Arabidopsis thaliana: a comparison to Homo sapiens B-ZIP motifs. Nucleic Acids Res 2004; 32(11):3435-3445.
35. Fong JH, Keating AE, Singh M. Predicting specificity in bZIP coiled-coil protein interactions. Genome Biol 2004; 5(2):R11.

36. Deppmann CD, Alvania RS, Taparowsky EJ. Cross-species annotation of basic leucine zipper factor interactions: Insight into the evolution of closed interaction networks. Mol Biol Evol 2006; 23(8):1480-1492.
37. Acharya A, Ruvinov SB, Gal J et al. A heterodimerizing leucine zipper coiled coil system for examining the specificity of a position interactions: Amino acids I, V, L, N, A and K. Biochemistry 2002; 41(48):14122-14131.
38. Krylov D, Mikhailenko I, Vinson C. A thermodynamic scale for leucine zipper stability and dimerization specificity: E and g interhelical interactions. EMBO J 1994; 13(12):2849-2861.
39. O'Shea EK, Rutkowski R, Kim PS. Mechanism of specificity in the Fos-Jun oncoprotein heterodimer. Cell 1992; 68(4):699-708.
40. Glover JN, Harrison SC. Crystal structure of the heterodimeric bZIP transcription factor c-Fos-c-Jun bound to DNA. Nature 1995; 373(6511):257-261.
41. Chen L, Glover JN, Hogan PG et al. Structure of the DNA binding domains from NFAT, Fos and Jun bound specifically to DNA. Nature 1998; 392(6671):42-48.
42. Messenguy F, Dubois E. Role of MADS box proteins and their cofactors in combinatorial control of gene expression and cell development. Gene 2003; 316:1-21.
43. Laughon A. DNA binding specificity of homeodomains. Biochemistry 1991; 30(48):11357-11367.
44. Li T, Stark MR, Johnson AD et al. Crystal structure of the MATa1/MAT alpha 2 homeodomain heterodimer bound to DNA. Science 1995; 270(5234):262-269.
45. Keleher CA, Goutte C, Johnson AD. The yeast cell-type-specific repressor alpha 2 acts cooperatively with a noncell-type-specific protein. Cell 1988; 53(6):927-936.
46. Tan S, Richmond TJ. Crystal structure of the yeast MATalpha2/MCM1/DNA ternary complex. Nature 1998; 391(6668):660-666.
47. Goutte C, Johnson AD. Recognition of a DNA operator by a dimer composed of two different homeodomain proteins. EMBO J 1994; 13(6):1434-1442.
48. Peterson BR, Sun LJ, Verdine GL. A critical arginine residue mediates cooperativity in the contact interface between transcription factors NFAT and AP-1. Proc Natl Acad Sci USA. 1996; 93(24):13671-13676.
49. Rao A, Luo C, Hogan PG. Transcription factors of the NFAT family: regulation and function. Annu Rev Immunol 1997; 15:707-747.
50. Lagrange T, Kapanidis AN, Tang H et al. New core promoter element in RNA polymerase II-dependent transcription: sequence-specific DNA binding by transcription factor IIB. Genes Dev 1998; 12(1):34-44.
51. Thomas MC, Chiang CM. The general transcription machinery and general cofactors. Crit Rev Biochem Mol Biol 2006; 41(3):105-178.
52. Thompson CC, Brown TA, McKnight SL. Convergence of Ets- and notch-related structural motifs in a heteromeric DNA binding complex. Science 1991; 253(5021):762-768.
53. Virbasius JV, Virbasius CA, Scarpulla RC. Identity of GABP with NRF-2, a multisubunit activator of cytochrome oxidase expression, reveals a cellular role for an ETS domain activator of viral promoters. Genes Dev 1993; 7(3):380-392.
54. Wolberger C. Combinatorial transcription factors. Curr Opin Genet Dev 1998; 8(5):552-559.
55. Tahirov TH, Inoue-Bungo T, Morii H et al. Structural analyses of DNA recognition by the AML1/Runx-1 Runt domain and its allosteric control by CBFbeta. Cell 2001; 104(5):755-767.
56. Bravo J, Li Z, Speck NA et al. The leukemia-associated AML1 (Runx1)—CBF beta complex functions as a DNA-induced molecular clamp. Nat Struct Biol 2001; 8(4):371-378.
57. Ghosh S, Gifford AM, Riviere LR et al. Cloning of the p50 DNA binding subunit of NF-kappa B: homology to rel and dorsal. Cell 1990; 62(5):1019-1029.
58. Muller CW, Rey FA, Sodeoka M et al. Structure of the NF-kappa B p50 homodimer bound to DNA. Nature 1995; 373(6512):311-317.
59. Ghosh G, van Duyne G, Ghosh S et al. Structure of NF-kappa B p50 homodimer bound to a kappa B site. Nature 1995; 373(6512):303-310.
60. Huang DB, Vu D, Cassiday LA et al. Crystal structure of NF-kappaB (p50)2 complexed to a high-affinity RNA aptamer. Proc Natl Acad Sci USA 2003; 100(16):9268-9273.
61. Chen YQ, Ghosh S, Ghosh G. A novel DNA recognition mode by the NF-kappa B p65 homodimer. Nat Struct Biol 1998; 5(1):67-73.
62. Brierley MM, Fish EN. Stats: multifaceted regulators of transcription. J Interferon Cytokine Res 2005; 25(12):733-744.
63. Fu XY, Kessler DS, Veals SA et al. ISGF3, the transcriptional activator induced by interferon alpha, consists of multiple interacting polypeptide chains. Proc Natl Acad Sci USA 1990; 87(21):8555-8559.
64. Angel P, Imagawa M, Chiu R et al. Phorbol ester-inducible genes contain a common cis element recognized by a TPA-modulated trans-acting factor. Cell 1987; 49(6):729-739.
65. Montminy MR, Sevarino KA, Wagner JA et al. Identification of a cyclic-AMP-responsive element within the rat somatostatin gene. Proc Natl Acad Sci USA. 1986; 83(18):6682-6686.

66. van Dam H, Huguier S, Kooistra K et al. Autocrine growth and anchorage independence: two complementing Jun-controlled genetic programs of cellular transformation. Genes Dev 1998; 12(8):1227-1239.
67. Kemler I, Schreiber E, Muller MM et al. Octamer transcription factors bind to two different sequence motifs of the immunoglobulin heavy chain promoter. EMBO J 1989; 8(7):2001-2008.
68. Tomilin A, Remenyi A, Lins K et al. Synergism with the coactivator OBF-1 (OCA-B, BOB-1) is mediated by a specific POU dimer configuration. Cell 2000; 103(6):853-864.
69. Botquin V, Hess H, Fuhrmann G et al. New POU dimer configuration mediates antagonistic control of an osteopontin preimplantation enhancer by Oct-4 and Sox-2. Genes Dev 1998; 12(13):2073-2090.
70. Dang CV, Barrett J, Villa-Garcia M et al. Intracellular leucine zipper interactions suggest c-Myc hetero-oligomerization. Mol Cell Biol 1991; 11(2):954-962.
71. Blackwood EM, Eisenman RN. Max: a helix-loop-helix zipper protein that forms a sequence-specific DNA-binding complex with Myc. Science 1991; 251(4998):1211-1217.
72. Prendergast GC, Lawe D, Ziff EB. Association of Myn, the murine homolog of max, with c-Myc stimulates methylation-sensitive DNA binding and ras cotransformation. Cell 1991; 65(3):395-407.
73. McMahon SB, Van Buskirk HA, Dugan KA et al. The novel ATM-related protein TRRAP is an essential cofactor for the c-Myc and E2F oncoproteins. Cell 1998; 94(3):363-374.
74. Eilers M, Schirm S, Bishop JM. The MYC protein activates transcription of the alpha-prothymosin gene. EMBO J 1991; 10(1):133-141.
75. Benvenisty N, Leder A, Kuo A et al. An embryonically expressed gene is a target for c-Myc regulation via the c-Myc-binding sequence. Genes Dev 1992; 6(12B):2513-2523.
76. Ayer DE, Lawrence QA, Eisenman RN. Mad-Max transcriptional repression is mediated by ternary complex formation with mammalian homologs of yeast repressor Sin3. Cell 1995; 80(5):767-776.
77. Ayer DE, Eisenman RN. A switch from Myc:Max to Mad:Max heterocomplexes accompanies monocyte/macrophage differentiation. Genes Dev 1993; 7(11):2110-2119.
78. Ayer DE, Kretzner L, Eisenman RN. Mad: A heterodimeric partner for Max that antagonizes Myc transcriptional activity. Cell 1993; 72(2):211-222.
79. Dang CV. c-Myc target genes involved in cell growth, apoptosis and metabolism. Mol Cell Biol 1999; 19(1):1-11.
80. Luscher B. Function and regulation of the transcription factors of the Myc/Max/Mad network. Gene 2001; 277(1-2):1-14.
81. Bouchard C, Dittrich O, Kiermaier A et al. Regulation of cyclin D2 gene expression by the Myc/Max/Mad network: Myc-dependent TRRAP recruitment and histone acetylation at the cyclin D2 promoter. Genes Dev 2001; 15(16):2042-2047.
82. Ron D, Habener JF. CHOP, a novel developmentally regulated nuclear protein that dimerizes with transcription factors C/EBP and LAP and functions as a dominant-negative inhibitor of gene transcription. Genes Dev 1992; 6(3):439-453.
83. Oyadomari S, Mori M. Roles of CHOP/GADD153 in endoplasmic reticulum stress. Cell Death Differ 2004; 11(4):381-389.
84. Batchvarova N, Wang XZ, Ron D. Inhibition of adipogenesis by the stress-induced protein CHOP (Gadd153). EMBO J 1995; 14(19):4654-4661.
85. Tang QQ, Lane MD. Role of C/EBP homologous protein (CHOP-10) in the programmed activation of CCAAT/enhancer-binding protein-beta during adipogenesis. Proc Natl Acad Sci USA 2000; 97(23):12446-12450.
86. Huang H, Lane MD, Tang QQ. Effect of serum on the down-regulation of CHOP-10 during differentiation of 3T3-L1 preadipocytes. Biochem Biophys Res Commun 2005; 338(2):1185-1188.
87. Benezra R, Davis RL, Lockshon D et al. The protein Id: A negative regulator of helix-loop-helix DNA binding proteins. Cell 1990; 61(1):49-59.
88. Ying QL, Nichols J, Chambers I et al. BMP induction of Id proteins suppresses differentiation and sustains embryonic stem cell self-renewal in collaboration with STAT3. Cell 2003; 115(3):281-292.
89. Perk J, Iavarone A, Benezra R. Id family of helix-loop-helix proteins in cancer. Nat Rev Cancer 2005; 5(8):603-614.
90. Iavarone A, Lasorella A. ID proteins as targets in cancer and tools in neurobiology. Trends Mol Med 2006; 12(12):588-594.
91. Ruzinova MB, Benezra R. Id proteins in development, cell cycle and cancer. Trends Cell Biol 2003; 13(8):410-418.
92. Liu D, Ishima R, Tong KI et al. Solution structure of a TBP-TAF(II)230 complex: Protein mimicry of the minor groove surface of the TATA box unwound by TBP. Cell 1998; 94(5):573-583.
93. Pereira LA, van der Knaap JA, van den Boom V et al. TAF(II)170 interacts with the concave surface of TATA-binding protein to inhibit its DNA binding activity. Mol Cell Biol 2001; 21(21):7523-7534.
94. Kokubo T, Swanson MJ, Nishikawa JI et al. The yeast TAF145 inhibitory domain and TFIIA competitively bind to TATA-binding protein. Mol Cell Biol 1998; 18(2):1003-1012.

95. Chicca JJ 2nd, Auble DT, Pugh BF. Cloning and biochemical characterization of TAF-172, a human homolog of yeast Mot1. Mol Cell Biol 1998; 18(3):1701-1710.
96. Kamada K, Shu F, Chen H et al. Crystal structure of negative cofactor 2 recognizing the TBP-DNA transcription complex. Cell 2001; 106(1):71-81.
97. Virbasius CM, Wagner S, Green MR. A human nuclear-localized chaperone that regulates dimerization, DNA binding and transcriptional activity of bZIP proteins. Mol Cell 1999; 4(2):219-228.
98. Matuoka K, Yu Chen K. Nuclear factor Y (NF-Y) and cellular senescence. Exp Cell Res 1999; 253(2):365-371.
99. Liberati C, di Silvio A, Ottolenghi S et al. NF-Y binding to twin CCAAT boxes: Role of Q-rich domains and histone fold helices. J Mol Biol 1999; 285(4):1441-1455.
100. Bellorini M, Lee DK, Dantonel JC et al. CCAAT binding NF-Y-TBP interactions: NF-YB and NF-YC require short domains adjacent to their histone fold motifs for association with TBP basic residues. Nucleic Acids Res 1997; 25(11):2174-2181.
101. Sorger PK. Heat shock factor and the heat shock response. Cell 1991; 65(3):363-366.
102. Liu PC, Thiele DJ. Modulation of human heat shock factor trimerization by the linker domain. J Biol Chem 1999; 274(24):17219-17225.
103. Wu C. Heat shock transcription factors: Structure and regulation. Annu Rev Cell Dev Biol 1995; 11:441-469.

CHAPTER 8

OLIGOMERIZATION AT THE MEMBRANE

Potassium Channel Structure and Function

Oliver B. Clarke and Jacqueline M. Gulbis*

The Walter and Eliza Hall Institute of Medical Research, Melbourne, Victoria, Australia; and Department of Medical Biology, The University of Melbourne, Melbourne, Victoria, Australia
Corresponding Author: Jacqueline M. Gulbis—Email: jgulbis@wehi.edu.au

Abstract: Cell membranes present a naturally impervious barrier to aqueous solutes, such that the physiochemical environment on either side of the lipid bilayer can substantially differ. Integral membrane proteins are embedded in this heterogeneous lipid environment, wherein the juxtaposition of apolar and polar molecular surfaces defines factors such as transverse orientation, the surface area available for oligomerisation and the symmetry of resultant assemblies. This chapter focuses on potassium channels –representative molecular pores that play a critical role in electrical signalling by enabling selective transport of K^+ ions across cell membranes. Oligomerization is central to K^+ channel action; individual subunits are nonfunctional and conduction, selectivity and gating involve manipulation of the common subunit interface of the tetramer. Regulation of channel activity can be viewed from the perspective that the pore of K^+ channels has coopted other proteins, utilizing a process of hetero-oligomerisation to absorb new functions that both enable the pore to respond to extrinsic signals and provide an electrical signature.

INTRODUCTION

Integral membrane proteins exist in a unique physicochemical and geometrical environment. The disposition of lipophilic and polar faces of the molecular surface of membrane proteins limits the possible orientations the protein can adopt, such that rotation is only permissible normal to the plane of the lipid bilayer or within a limited angular range. Moreover, as individual membrane proteins rotate, translate and interact with one another within a geometrically constrained framework, the possibilities for

Protein Dimerization and Oligomerization in Biology, edited by Jacqueline M. Matthews.

oligomerisation symmetry are also restricted. In homomeric membrane assemblies only cyclically symmetric, or slightly asymmetric permutations, are achievable, whereas amongst soluble homomeric complexes, dihedrally symmetric complexes are more common than cyclic complexes comprising the same number of subunits.[1] Baker and coworkers recently showed (for soluble proteins) that symmetric complexes are generally lower in energy than their asymmetric counterparts with the same number of subunits and hence that they dominate the population of low-energy assemblies available for modification by natural selection.[2] It is likely that this also holds true for membrane proteins.

The physiochemical environment of the membrane is uniquely heterogeneous. In integral membrane proteins transmembrane elements of the protein are located in the hydrophobic interior of the lipid bilayer, whilst regions of the protein on either side of the membrane are exposed to environments that may be substantially out of equilibrium with one another, in terms of redox potential, electrochemical potential and the divergent populations of chemical species that reside there. The outer and inner surfaces of the lipid bilayer are further determinants of the environment occupied by an integral membrane protein, having substantially different chemical compositions in the inner and outer leaflets.

On a molecular level, membrane protein oligomers may be segregated into three classes upon the basis of structure and function. In the first class are proteins for which oligomerisation is an obligate state, commencing at cotranslation and irreversible over the lifetime of the membrane protein. Ion channels are an example of this class and they form the primary focus of this chapter. The second class contains proteins that form transient oligomers during signal transduction—of which cytokine receptors provide a clear example. The third class entails oligomerisation of mature proteins, perhaps as a consequence of molecular signalling, but for which oligomer assembly is irreversible once completed. Examples of this class are the cytolytic membrane attack complex that forms during the humoural immune response to pathogen invasion and the pro-apoptotic regulators of the Bcl-2 family, Bax and Bak.

A feature typifying many membrane proteins assemblies is a continuous channel spanning the membrane, allowing selective transport or diffusion of particular chemical species. This chapter focuses on ion channels, specifically potassium channels, molecular pores that enable the selective and passive transport of potassium ions across the membrane and have the structural and functional diversity to inform many aspects of oligomerisation in membrane biology.

Oligomerisation serves several purposes in channel biology. It facilitates the assembly of complex molecular machinery in a manner that is highly economical with genetic information. The modular nature of ion channel assembly also allows tuning of functional properties; incorporation of closely related channel monomers or alternatively spliced variants form subtly different heteromers that optimize functional characteristics for a particular situation. In addition, the quaternary structure of ion channels is intimately related to the nature and symmetry of the substrate. The 4-fold symmetry of the potassium channel, for example, provides 8-fold coordination selective for potassium ions that transit through the selectivity filter.

POTASSIUM CHANNELS

This chapter uses specific examples to illustrate how oligomerisation of potassium channels pertains to almost every aspect of their biology, from ion conduction to the fine-tuning of responsiveness. Over several decades, electrophysiological studies have revealed K^+ channels to be highly complex molecular systems that effectively modulate the cell potential in response to extracellular or intracellular signals. By forming a physical barrier to bulk lipid, the helical transmembrane segments of these channels counter a prospective high energetic penalty to transport of charged particles across electrically neutral cell membranes, thereby satisfying the primary requirement for enabling a controlled flow of ions. Potassium channels are obligate tetramers (or in some instances pseudo-tetrameric dimers)[3] and the central ion permeation pathway is consequently characterized by 4-fold rotational symmetry. The symmetry of the pore is echoed in regulatory domains that are responsive to extrinsic signals, where the quadrupling of sensors potentially determines the sensitivity of effector binding and response via subunit cooperativity.

Potassium channels share common pore architecture but take on almost countless forms. Major families are classified according to electrical signature and regulatory signals to which they respond. Voltage-gated Kv channels, for example, respond to waves of membrane depolarisation propagated during neuronal transmission, whereas large conductance BK (Maxi-K; *Slopoke*) channels respond to a combination of intracellular Ca^{2+} levels and voltage. Inward rectifiers typically respond to phosphatidyl inositides and closely associated partner proteins and are characterized by an ability to limit ion efflux. The natural incidence of particular K^+ channels is further specified by the organ or tissue, or appearance at a particular stage in cell differentiation. Potassium channel families are conserved throughout the evolutionary tree, even in unicellular organisms, and their structural diversity reflects a multitude of functions. In multicellular organisms, K^+ currents are essential to electrical signalling, governing central nervous system, cardiac, renal and a host of other organ functions. As the evolutionary tree is scaled, the range of complex electrical signalling requirements increases. In the central nervous system of mammals, for example, K^+ channels set the resting potential of cells and modulate the duration and frequency of action potentials[4] and in a beating heart they help to achieve and maintain precise rhythmic firing. In each organ, nuance within its electrical signature may entail interplay between different types of ion channel. The critical role of K^+ channels in organ function has elevated their status as therapeutic targets.

Architecturally, K^+ channel subunits have antiparallel outer and inner membrane-spanning helices and a shorter pore helix that spans only the outer leaflet of the bilayer.[5] An amphiphilic N-terminal extension of the outer helix at the intracellular surface of the membrane is termed the slide helix.[6] Three major conserved features of K^+ channel action specific to the transmembrane pore are ion selectivity, conduction and gating. All of these functions localize to an ion permeation pathway coincident with the molecular axis at the common interface of the subunits.

Ion Selectivity

A narrow filter region in the outer leaflet of the bilayer confers ion selectivity. The molecular scaffold precisely positions four symmetry-related linkers containing the K^+ channel consensus motif (T, X, G, Y/F, G).[7] The C-terminal end of each pore helix anchors the N-terminal end of the linker, while its other end connects to the extracellular face of the channel. Thus four consensus linkers are precisely aligned to optimize coordination of K^+ in preference to other cations.[8,9]

In the selectivity filter, the two consensus glycine residues permit free rotation about the main chain amides and consequently the orientation of the backbone carbonyls is virtually unrestricted. In crystal structures of K^+ channels, however, all of the carbonyls in the selectivity filter face inward in approximate alignment.[10] Their orientation is determined by a string of central K^+ ions, coordinated by parallel rings of carbonyls at discrete sites along the selectivity filter. The K^+ ions are an integral part of the molecular structure of the channel and the carbonyl oxygens provide an effective substitute for hydrating water molecules stripped from each K^+ ion as it enters the pore.

Conduction

The selectivity filter governs conduction as well as selectivity and must be capable of deformation and recovery on a microsecond timescale[11] to accommodate approximately ten million ions per second (a rate approaching the natural diffusion limit). It is widely held that electrostatic repulsion between ions at neighbouring coordination sites renders simultaneous occupation unfavourable. Charge repulsion is thought to provide the impetus for an otherwise passive process of conduction down an electrochemical gradient. Although a K^+ channel could conceivably be composed of a symmetric trimer, as in the β-barrel efflux channel TolC,[12] or pentamer as in the mechanosensitive channel MscL,[13] only tetramers have been observed. The tetrameric geometry provides a rationale for the co-existence of transience and selectivity, in that it favours symmetric coordination of each K^+ by eight backbone carbonyls, rather than by the preferred K^+ coordination number of seven, and the small energy penalty incurred may be sufficient to prevent stable binding.[9]

Gating

Gating is a complex issue. Ion conduction requires a clear path through the membrane and is contingent upon gating elements in the permeation pathway adopting an "open" status, implicating structural changes or perturbations at the tetramer interface. Considerable functional evidence indicates that a gate is located at the inner-helix bundle at the face of the membrane.[14-21] This has been termed the activation or intracellular gate and is thought to govern K^+ access by adopting discrete closed or open conformations. Moreover, research is increasingly focused on the possibility that the selectivity filter also acts as a gate.[22-29] Whilst the roles and interdependency of the conduction gates have not yet been fully elucidated, it should be noted that 4-fold symmetry of the tetramer interface increases the susceptibility of the permeation pathway to global conformational change.

PORE SYMMETRY

The symmetry of the K^+ channel pore occurs at the level of secondary and tertiary structure, not necessarily reflected in the primary sequence. Obligate oligomerisation allows some K^+ channels to heteromerize from sibling members of a family,[30] providing a means of conferring subtle changes and a range of variation in the electrical profile, depending on the constitution of the pore. Members of the Kv1 family, for example, are known to co-assemble into heteromeric channels.

Although there are as yet no crystal structures of Na^+ or Ca^{2+} channels, they are expected to have a similar topology to K^+ channels and indeed, database mining has identified homotetrameric Na^+ channels in prokaryotes. In higher organisms, pseudo-tetramers arise from a single polypeptide chain containing four inexact repeats, suggesting a lack of uniformity about the molecular tetrad for reasons as yet unknown. The reasons may be manifold and one possibility is that domain cooperativity in Na^+ channels speeds up gating,[31] congruent with the role of Na^+ channels in generating the rising phase of action potentials, rather than the slower recovery phase mediated by K^+ channels. Another is that a level of asymmetry permits fine-tuning of the coordination geometry in the selectivity filter to improve ion selectivity in multicellular organisms. A third possibility is that the difference in assembly format ensures Na^+ and K^+ subunits never accidentally co-assemble.

MORPHOLOGY OF THE PORE

Although the secondary structure and topology is conserved, crystal structures of K^+ channels from different families have revealed marked conformational plasticity of the canonical pore structure. Other integral membrane proteins (e.g., the Na^+/H^+ antiporter NhaA[32]) have been crystallized in alternate conformations, which have been reported as compatible with discrete functional states. Whilst it has not been verified that significant conformational change of any individual species of K^+ channel occurs, the conformational variation observed *across* K^+ channel families has been attributed to physiological status. In K^+ channels, the major apparent differences correspond to the conformation of the inner helices, affecting the width of an inner vestibule in the permeation pathway between the mid-point and the intracellular face of the membrane. In structures of Ca^{2+}-gated and voltage-gated K^+ channels, the inner helices are bent internally at conserved glycine or proline residues,[33,34] resulting in a comparatively wide intracellular entrance to the conduction pathway such that the vestibule is continuous with the intracellular solution. In reported structures of KcsA and inward rectifiers,[5,6] however, the inner helices are relatively straight. On this basis, structures of Ca^{2+} and voltage-gated K^+ channels have been assigned as having an open status, in which the central permeation pathway is readily accessible to ions and KcsA and inward rectifier structures assigned to closed status, where ion conduction is blocked. An assumption implicit in this line of reasoning is that K^+ channels switch between alternate states in response to a gating cue and that the inner helices are subject to lateral hinge motions about the conserved glycine or proline residues.

REGULATORY ASSEMBLIES

Potassium channel families are distinguishable on the basis of regulatory assemblies that gate and modulate current, providing a vital link between electrical activity and cellular chemistry. Voltage-gated K^+ *Shaker*-like (Kv1) channels, for example, have both regulatory transmembrane domains, intracellular adaptor domains that determine specificity of assembly[35,36] and auxiliary β-subunits. These three parts co-assemble into a large quaternary complex that shares the 4-fold symmetry of the pore.[34,37] Conservation of 4-fold symmetry is not applicable to all K^+ channels in all gating states, however, as evidenced by the recent structure of full-length KcsA,[38] which exhibits overt 2-fold symmetry of a sizeable intracellular assembly, compatible with, but distinct from, the pore.

Gating domains are encoded in the same open reading frame as the channel or as auxiliary subunits (typically termed β-subunits) encoded in separate genes that are cotranslated with the channel.[39,40] It has been shown that surface expression and trafficking of Kv1 (*Shaker*-related) K^+ channels is dependent on prior assembly with Kvβ-subunits,[41,39] with evidence from the large conductance Maxi-K family,[42] the small conductance KCNQ channels[43] and the α-dendrotoxin-sensitive channels,[44] amongst others, suggesting that this is not a unique situation. In multi-domain K^+ channels, the proximity of N- and C-terminal regions extrinsic to the pore facilitates close interactions in the quaternary structure. Moreover, structures of representative inward rectifier[6] and voltage-gated[34] K^+ channels have respectively revealed subtle and conspicuous entwining of adjacent subunits in a manner akin to domain swapping, in each instance serially connecting the four subunits into a circle. The following sections describe examples of channel hetero-oligomerisation involving regulatory domains and partner proteins.

K^+ CHANNELS WITH ADDITIONAL DOMAINS OR SUBUNITS WITHIN THE MEMBRANE

Voltage-Dependent K^+ Channels

Voltage-gated channels respond to changes in the membrane potential. Their ability to respond is conferred by specialized voltage sensing domains attuned to the electric field of the membrane. The voltage sensor has four transmembrane segments (conventionally S1 to S4) and is N-terminal to the pore (S5 and S6) (Fig. 1A). A helical segment linking the two regions is akin to the slide helix in Kir channels. Whilst it has been shown that structural changes involving the voltage sensors coincide with the relay of sensory input to a gate in the permeation pathway,[45,46] the nature and magnitude of the changes and the means by which they activate the gate, remain unresolved.

The voltage sensor contains a consensus motif, wherein the conserved sequence corresponds to regular punctuation of S4 by an arginine (or lysine) at every third residue. The ability of charged side chains within a membrane to reorient the electric field implicated the arginines as candidates for voltage sensing.[47-49] Similar motifs are present in other types of voltage-gated cation channels (e.g., sodium channels), suggesting that voltage sensing predates selectivity in ion channel evolution. Under appropriate experimental conditions,[50,51] a 'gating current' can be measured upon membrane depolarisation. The gating current is distinguishable from the ion current by its magnitude and in Kv channels equates to approximately thirteen charges crossing the entire transmembrane voltage difference.[52,53] The gating current of outward positive charge movement has been attributed to the consensus arginines,[53,54] which is consistent with marked changes in accessibility of the arginine residues in S4 and the S4-S5 linker to the intracellular and extracellular solutions during pore gating.[46] Underpinned by the assumption that the transmembrane voltage difference varies approximately linearly with distance through the bilayer, gating current and S4 accessibility data have lead to the widely accepted conclusion that the entire S4 segment moves outward and moreover that the mechanics of S4 motion govern the status at the activation gate. More recently, studies based on data from a variety of biophysical methods have produced alternative models differing in aspects of rotation and translation of S4, some of which take into consideration focusing of the electric field.

The first structure of a Kv family member, the prokaryotic channel KvAP,[55] exhibited clear-cut separation of voltage-sensor and pore, upholding the concept that distinct domains exist within the membrane and, moreover, inferring that significant relative movement between the two domains is possible. The disposition of the domains of

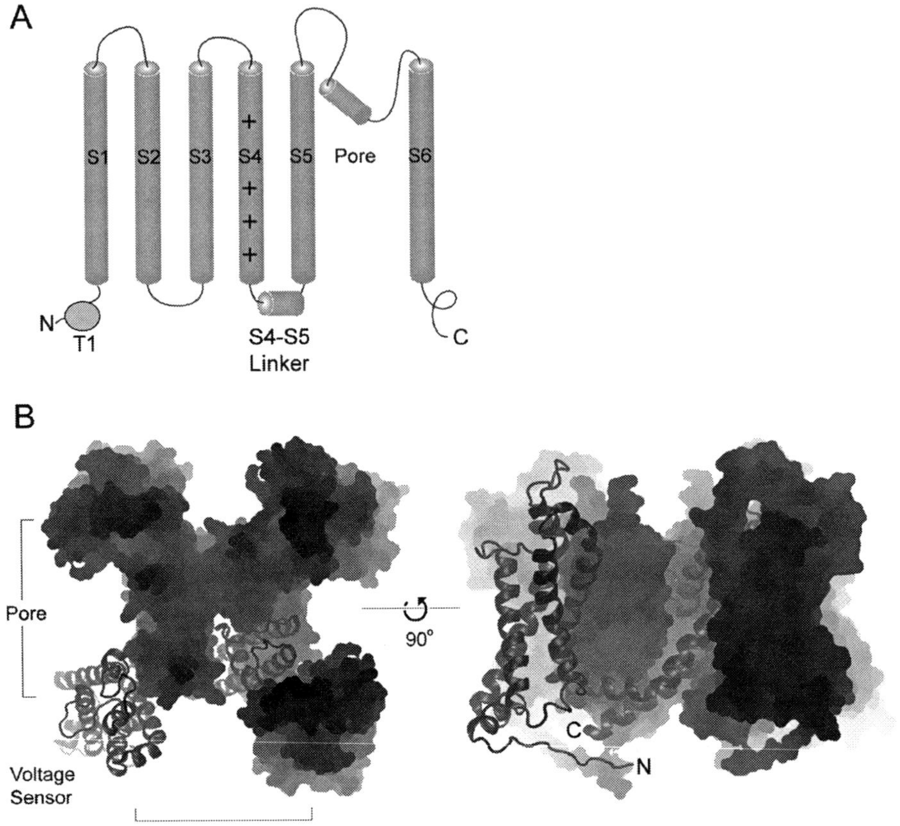

Figure 1. Voltage-gated Kv channels have two domains within the membrane. A) A schematic of the primary sequence. Four helical transmembrane segments N-terminal to the pore (S1 to S4) form a voltage-sensing domain. The N-terminus forms a cytoplasmic domain of approximately 100 residues known as the tetramerisation, or T1, domain. B) A surface representation of the transmembrane regions of Kv1.2 with one subunit represented as a ribbon is viewed from the extracellular side (left) and in the plane of the membrane (right). The ribbon shows that the two transmembrane helices comprising the pore do not interface with the directly abutting voltage sensor, but interweave such that they form an extensive interface with an adjacent subunit.

KvAP did not altogether make sense in the context of the lipid bilayer, however and it is now generally accepted that the structure represents a nonphysiological conformation due to membrane extraction, purification and/or crystallisation artefact. The more recent structure of the entire assembly of a mammalian voltage-dependent K^+ channel (Kv1.2), including a large intracellular assembly comprising N-terminal domains and β-subunits,[34] unexpectedly revealed substantial intertwining of the membrane domains. In the crystal structure, both the S4-S5 linker and voltage sensor of each subunit form narrow but extensive interfaces with the pore domain of an interleaving subunit (Fig. 1B). The interfaces are mediated by aliphatic and aromatic side chain contacts and run the length of the helices, burying a significant area of the molecular surface within the interfacial regions (in Kv1.2 this is 1,668 Å^2 per subunit). In the absence of further structural information, it is difficult to envisage how substantial interdomain movement might occur in the light of the hydrophobic character of the interfaces interweaving the four subunits.

Intriguingly, the voltage sensor has a homologue in a phosphatase from *Ciona intestinalis* Ci-VSP, which is similarly voltage dependent and exhibits a gating current under voltage clamp.[56-58] It demonstrates the functional modularity of K^+ channels; in this instance that voltage-sensitivity exists independently of the pore, as well as vice versa. Two recently identified proton channels, HV1[59] and VSOP,[60] neither of which is associated with a distinct pore domain, are also homologues of the voltage sensor.

A mechanistic appreciation of these interdependent, coexistent, functionalities in a single protein has proved elusive and debate over voltage gating is ongoing. The focus is on S4 motion, as described earlier, and reported estimates of distances tracked by the S4 positive charges vary by an order of magnitude. A minimum value of around 2 Å 'vertical' displacement determined by luminescence energy transfer[61] is at odds with a value of 15-20 Å determined by biotin trapping experiments on KvAP.[55] The latter has been interpreted as a 'paddle'-like motion of the sensor relative to the pore. The use of voltage clamp fluorimetry in combination with molecular simulation methods suggested an intermediate distance of 6-8 Å and was the first study to propose conformational change *within* the voltage sensor.[62] Recent molecular dynamics analyses went further in proposing that zipper-like rearrangement of salt-bridges within the voltage sensor is responsible for the gating charge current,[63,64] as opposed to substantive movement of the entire helix. In order to reconcile vastly different standpoints, the discrepancies must be viewed in the light of experimental factors and limitations, such as how well each technique accounts for lateral versus vertical motion, the change in transmembrane voltage across the membrane and rotation of the sensor within the membrane. There are some excellent reviews on this topic.[65,66]

HETERO-OLIGOMERISATION: PARTNER PROTEINS IN THE MEMBRANE

K_{ATP} Channels

Co-assembled regulatory proteins represent an alternative to intrinsic sensors. The ATP-sensitive K^+ channel assemblies (K_{ATP}) are an example of partnering subunits acting as metabolic sensors.

The two components comprising K_{ATP} channels are cotranslated in the endoplasmic reticulum before being transported to the cell surface.[67] The functional assembly is octameric, with the permeation pathway of an inward rectifier K^+ channel (Kir) co-assembled with four sulfonylurea receptors (SUR). Sulfonylurea receptors interact exclusively with channels of the Kir6 subfamily (Kir6.1 and Kir6.2) and neither channel nor receptor functions independently in vivo. The sulfonylurea receptors are ATP-binding cassette (ABC) transporters that have lost an intrinsic solute transport function but retain effector and nucleotide-binding capacity. Three isoforms, SUR1, SUR2A and SUR2B, confer differing attributes to the K_{ATP} assemblies. For example, atrial and ventricular K_{ATP} channels have distinct pharmacological properties due to differential association of Kir6.2 with SUR1 or SUR2A respectively.[68] Whilst functional

ABC transporters have only two transmembrane domains (TMD1 and TMD2), an extra domain (TMD0) at the N-terminus of SURs mediates interaction with Kir6 members.[69,70] Each SUR subunit has a total of seventeen transmembrane helices (Fig. 2) and two intracellular nucleotide-binding domains (NBD) with differing nucleotide affinities and magnesium requirements.[71,72] Kir6 subunits for their part have a large intracellular domain implicated in regulation and current rectification.

K_{ATP} assemblies are found in excitable cells, notably including cardiac tissue [e.g., Weiss and Lamp, 1987] and pancreatic β cells, where they are responsible for insulin exocytosis.[73,74] They have also been implicated in physiological stress responses.[75-78] A well-known aspect of their function is in providing a link between pancreatic β-cell metabolism (ADP/ATP ratio) and electrical activity of the plasma membrane. Congenital defects lowering the sensitivity of the channel to inhibitory ATP result in overactive Kir6.2/SUR1 channels and give rise to neonatal diabetes mellitus.

In the absence of SUR subunits, wild-type Kir6.2 does not express functional channels. The problem does not appear to lie with the channel subunits, as C-terminally truncated Kir6.2 mutants (Kir6.2Δ26 and Kir6.2Δ36) exhibit currents comparable to those of the complete K_{ATP} assembly that are similarly inhibited by cellular ATP.[79] The loss of channel activity in membrane patches is most likely caused by a failure of trafficking and membrane insertion.[79,80] In wild-type channels, co-assembly of Kir6.2 with SUR1 confers the ability to conduct K^+ ions, at a price. The SUR1 subunits are able to effect pore opening and closing, coupling pore activation to the binding of metabolic or therapeutic effectors. Sulfonylurea inhibits channel activity from a site on the C-terminal transmembrane helix of the receptor. A concrete example of control over the pore by SUR1 is that neonatal diabetes is now treatable by therapeutic administration of sulfonylureas to bypass β-cell metabolism.[81-83]

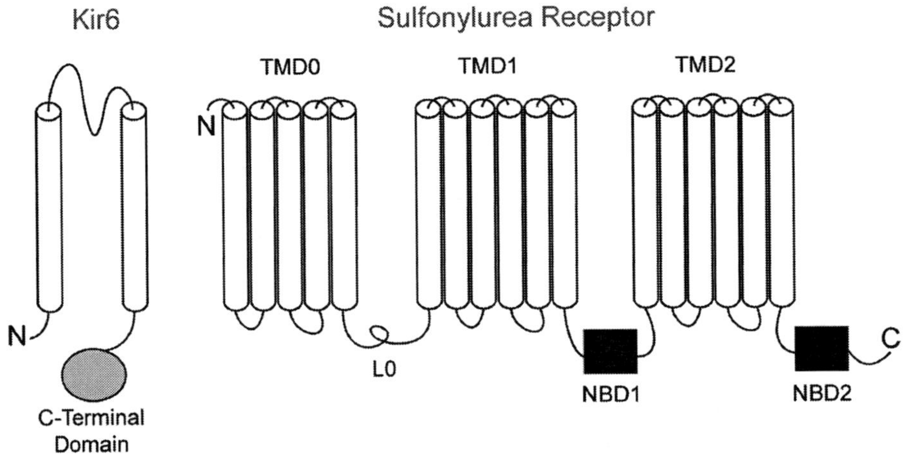

Figure 2. K_{ATP} channels are hetero-octamers. A schematic shows the relative size and complexity of pore and regulatory subunits. Each Kir6 subunit in the tetramer coassembles with an ABC transporter subunit. The TMD0 domain and L0 linker are modifications to the basic ABC transporter fold that form the primary connection to the pore.

INTRACELLULAR ASSEMBLIES

By transducing signals to the pore, specialized gating domains equip the channel to respond to intracellular agents and to adapt channel behaviour in stress conditions (e.g., pH, oxidation). The examples below are of subunits that co-assemble with the pore.

K_{ATP} Channels

The intracellular domains of Kir6.2 retain partial control of the pore.[84] Adenine nucleotides act on both channel and receptor subunits, with apparently non-overlapping modes of action. ATP bound to Kir6.2Δ26 or Kir6.2Δ36 stabilizes the K^+ channel in a closed state.[79] The ATP-binding site is putatively located near an intersubunit parallel strand interaction connecting the intracellular N- and C-terminal domains.[85,86] Conversely, in SUR subunits, Mg^{2+}/ADP (MgADP) stimulates action via the NDBs. The ADP and ADP occupy classic nucleotide-binding pockets defined by highly conserved Walker A and Walker B motifs and the characteristic glutamine (Q-loop) and histidine (H-loop) residues.

In order to understand the regulatory effects of oligomerisation, at least in terms of how MgADP acts on SUR subunits in K_{ATP} channels, it is instructive to consider the situation in functional ABC transporters. Prior to determination of a range of full-length ABC transporter structures, crystal structures of point mutant and native NBDs from *E. coli* Haemolysin B (HlyB) ABC transporter [Zaitseva et al, 2006] lent valuable clues as to how the NDBs might relay signals to the transmembrane domains of ABC transporters by showing the NBD in monomeric and dimeric states, with and without ATP, ADP and Mg^{2+}. Similarly, electron microscopy visualisation of the MDR1 ABC transporter indicated that large structural rearrangements were possible [Rosenberg MF et al, 2003]. Differential rotations of catalytic and helical domains in the HlyB NDB structures identified a molecular hinge region that includes the Q-loop, providing a route for long-range transmission of local ATP chemistry to the membrane. Crystal structures of full-length ABC transporters[87-89] showed juxtaposed NBDs and affirmed multiple conformational states.

In general, structural and biochemical evidence points to cycles of binding and hydrolysis as drivers of solute transport, harvesting the energy of concurrent conformational changes in the NBDs and transducing it to the transmembrane regions. The coupling mechanism is not yet understood with any certainty, although common threads in all posited models are that ATP binding, hydrolysis and release are linked to association or disassociation of NBD dimers and that hidden in the process is a 'power stroke'. Whilst in some ABC transporters this has been attributed to the nucleotide-binding step,[87] the means remains unclear for SUR subunits. It appears certain, however, that in K_{ATP} channels, SUR and Kir subunits are intimately coupled in such a way that the mechanical and chemical energy of the ATP catalytic cycle is co-opted to switch gating states of the K^+ pore.

Kv Beta Subunits

In the central nervous system of higher organisms, voltage-gated channels of the Kv1 and Kv4 families form an exclusive lifetime association with cytoplasmic proteins known as β-subunits[90,91] (the α-subunits comprise the pore). Co-translation of Kvα and Kvβ-subunits in the endoplasmic reticulum promotes N-linked glycosylation and membrane

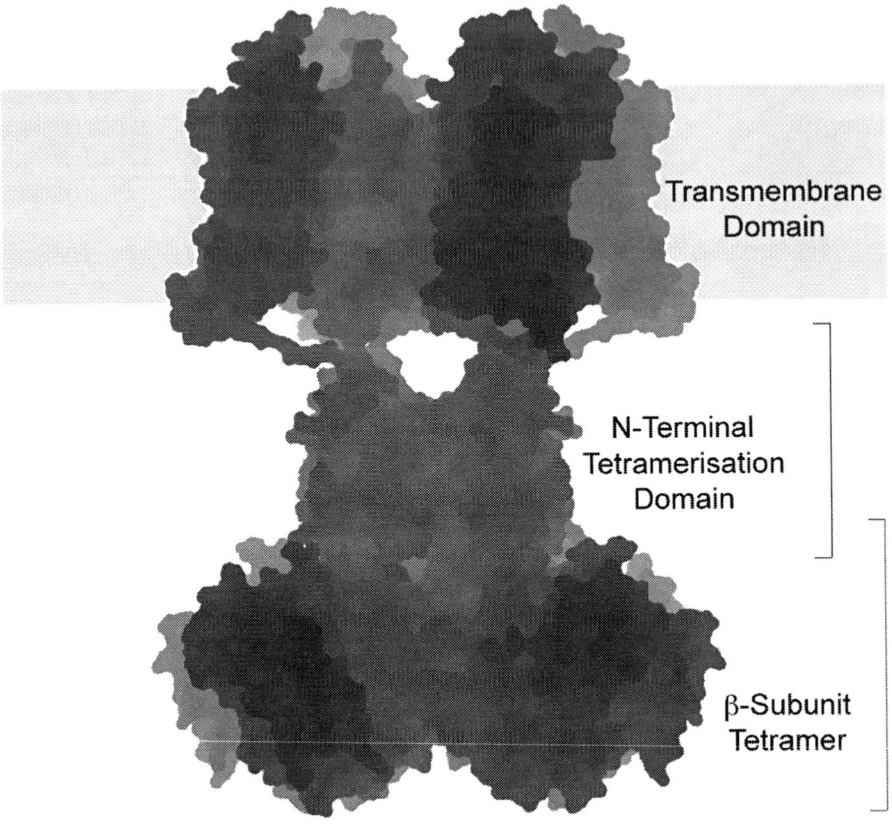

Figure 3. Kvβ subunits are located on the intracellular side of the membrane. A depiction of the molecular surface of Kv1.2 shows a large octameric assembly of the N-terminal tetramerisation (T1) domain and Kvβ subunits suspended beneath the pore and voltage sensor domains. The cytoplasmic assembly retains the 4-fold rotational symmetry of the pore.

trafficking of the channel,[41] ensuring that Kvβ is present on functional channels in the membrane. The Kvβ-subunits are members of the aldo-keto reductase (AKR) superfamily, retaining a catalytic centre and cofactor (NADP) binding. A role in regulation of the channel appears likely; if bound NADP is oxidized, both the open probability of the permeation pathway[92] and the magnitude of the current[93] exhibit an increase.

The large intracellular assembly of Kv1 consists of the N-terminal assembly of the channel α-subunits associated with one face of a symmetric tetramer of Kvβ subunits[37,94] (Fig. 3). Of the three known Kvβ isoforms, Kvβ1 and Kvβ3 have an N-terminal 'inactivation particle' that performs the auxiliary function of stemming current flow by blocking the permeation pathway.[95,96] The lack of an activation particle makes Kvβ2 a simpler system for electrophysiological and other analyses. Experiments have shown that Kvβ2 null mice and knock-in mice with catalytic point mutants of Kvβ2 show similar symptoms to Kv1.1 null mice, including short life spans and seizures,[97,98] implying that proper functioning of Kv1 channels is reliant upon both α and β-subunits. Alongside functional data linking redox activity of Kvβ subunits to channel activity, the evidence supports a now general consensus that Kvβ subunits contribute to excitability in vivo, by atypical application of intrinsic AKR catalytic activity.[92,93,97,99,100]

CONCLUSION

Potassium channels exemplify functional aspects pertinent to molecular pores and transporters in cell membranes, in particular the adoption of discrete conformational states during activation and control over permeation by co-assembled subunits or discrete domains that respond to extrinsic signals. Subunit cooperativity during action is another factor that may influence the kinetics of response. Oligomerisation is essential for the major aspects of K^+ channel function, including conduction, selectivity, gating and regulation. A modular constitution enables the basic structural framework to be tuned and modified by selective assembly of pore and regulatory subunits from differing isoforms, adding nuance to the electrical signature of K^+ channels.

ACKNOWLEDGEMENTS

We thank Dr Brian Smith at The Walter and Eliza Hall Institute for help with figure preparation and insightful comments.

REFERENCES

1. Levy ED, Boeri Erba E, Robinson CV et al. Assembly reflects evolution of protein complexes. Nature 2008; 453(7199):1262-5.
2. Andre I, Strauss CE, Kaplan DB et al. Emergence of symmetry in homooligomeric biological assemblies. Proc Natl Acad Sci USA 2008; 105(42):16148-52.
3. Patel AJ, Honore E, Maingret F et al. A mammalian two pore domain mechano-gated S-like K+ channel. EMBO J 1998; 17(15):4283-90.
4. Hille B. Ionic Channels of Excitable Membranes. 2nd Edition ed. Sunderland, Massachusetts: Sinauer Associates Inc.; 1992.
5. Doyle DA, Morais Cabral J, Pfuetzner RA et al. The structure of the potassium channel: molecular basis of K+ conduction and selectivity. Science 1998; 280(5360):69-77.
6. Kuo A, Gulbis JM, Antcliff JF et al. Crystal structure of the potassium channel KirBac1.1 in the closed state. Science 2003; 300(5627):1922-6.
7. Heginbotham L, Lu Z, Abramson T et al. Mutations in the K+ channel signature sequence. Biophys J 1994; 66(4):1061-7.
8. Bernèche S, Roux B. Energetics of ion conduction through the K+ channel. Nature 2001; 414(6859):73-7.
9. Bostick DL, Brooks CL. Selectivity in K+ channels is due to topological control of the permeant ion's coordinated state. Proc Natl Acad Sci USA 2007; 104(22):9260-5.
10. Zhou Y, Morais-Cabral JH, Kaufman A et al. Chemistry of ion coordination and hydration revealed by a K+ channel-Fab complex at 2.0 A resolution. Nature 2001; 414(6859):43-8.
11. Berneche S, Roux B. Molecular dynamics of the KcsA K+ channel in a bilayer membrane. Biophysical Journal 2000.
12. Koronakis V, Sharff A, Koronakis E et al. Crystal structure of the bacterial membrane protein TolC central to multidrug efflux and protein export. Nature 2000;405(6789):914-9.
13. Chang G, Spencer RH, Lee AT et al. Structure of the MscL homolog from Mycobacterium tuberculosis: a gated mechanosensitive ion channel. Science 1998;282(5397):2220-6.
14. Armstrong CM. Inactivation of the potassium conductance and related phenomena caused by quaternary ammonium ion injection in squid axons. J Gen Physiol 1969; 54(5):553-75.

15. Armstrong CM. Interaction of tetraethylammonium ion derivatives with the potassium channels of giant axons. J Gen Physiol 1971; 58(4):413-37.
16. Armstrong CM, Hille B. The inner quaternary ammonium ion receptor in potassium channels of the node of Ranvier. J Gen Physiol 1972; 59(4):388-400.
17. Miller C. Trapping single ions inside single ion channels. Biophys J 1987; 52(1):123-6.
18. MacKinnon R, Yellen G. Mutations affecting TEA blockade and ion permeation in voltage-activated K+ channels. Science 1990; 250(4978):276-9.
19. Yellen G, Jurman ME, Abramson T et al. Mutations affecting internal TEA blockade identify the probable pore-forming region of a K+ channel. Science 1991; 251(4996):939-42.
20. Heginbotham L, Abramson T, MacKinnon R. A functional connection between the pores of distantly related ion channels as revealed by mutant K+ channels. Science 1992; 258(5085):1152-5.
21. Baukrowitz T, Yellen G. Use-dependent blockers and exit rate of the last ion from the multi-ion pore of a K+ channel. Science 1996; 271(5249):653-6.
22. Hoshi T, Zagotta WN, Aldrich RW. Two types of inactivation in Shaker K+ channels: effects of alterations in the carboxy-terminal region. Neuron 1991; 7(4):547-56.
23. Lu T, Ting A, Mainland J et al. Probing ion permeation and gating in a K+ channel with backbone mutations in the selectivity filter. Nat Neurosci 2001; 4(3):239-46.
24. Cordero-Morales JF, Cuello LG, Zhao Y et al. Molecular determinants of gating at the potassium-channel selectivity filter. Nat Struct Mol Biol 2006; 13(4):311-8.
25. Cordero-Morales JF, Jogini V, Lewis A et al. Molecular driving forces determining potassium channel slow inactivation. Nat Struct Mol Biol 2007; 14(11):1062-9.
26. Blunck R, Cordero-Morales JF, Cuello LG et al. Detection of the opening of the bundle crossing in KcsA with fluorescence lifetime spectroscopy reveals the existence of two gates for ion conduction. The Journal of General Physiology 2006; 128(5):569-81.
27. Ader C, Schneider R, Hornig S et al. Coupling of activation and inactivation gate in a K+-channel: potassium and ligand sensitivity. EMBO J 2009; 28(18):2825-34.
28. Baukrowitz T, Yellen G. Modulation of K+ current by frequency and external [K+]: a tale of two inactivation mechanisms. Neuron 1995; 15(4):951-60.
29. Gao L, Mi X, Paajanen V et al. Activation-coupled inactivation in the bacterial potassium channel KcsA. Proc Natl Acad Sci USA 2005; 102(49):17630-5.
30. Po S, Roberds S, Snyders DJ et al. Heteromultimeric assembly of human potassium channels. Molecular basis of a transient outward current? Circ Res 1993; 72(6):1326-36.
31. Chanda B, Asamoah OK, Bezanilla F. Coupling interactions between voltage sensors of the sodium channel as revealed by site-specific measurements. J Gen Physiol 2004; 123(3):217-30.
32. Appel M, Hizlan D, Vinothkumar KR et al. Conformations of NhaA, the Na+/H+ exchanger from Escherichia coli, in the pH-activated and ion-translocating states. J Mol Biol 2009; 388(3):659-72.
33. Jiang Y, Lee A, Chen J et al. Crystal structure and mechanism of a calcium-gated potassium channel. Nature 2002; 417(6888):515-22.
34. Long SB, Campbell EB, Mackinnon R. Crystal structure of a mammalian voltage-dependent Shaker family K+ channel. Science 2005; 309(5736):897-903.
35. Shen NV, Chen X, Boyer MM et al. Deletion analysis of K+ channel assembly. Neuron 1993; 11(1):67-76.
36. Li M, Jan YN, Jan LY. Specification of subunit assembly by the hydrophilic amino-terminal domain of the Shaker potassium channel. Science 1992; 257(5074):1225-30.
37. Gulbis JM, Zhou M, Mann S et al. Structure of the cytoplasmic beta subunit-T1 assembly of voltage-dependent K+ channels. Science 2000; 289(5476):123-7.
38. Uysal S, Vasquez V, Tereshko V et al. Crystal structure of full-length KcsA in its closed conformation. Proc Natl Acad Sci USA 2009; 106(16):6644-9.
39. Nagaya N, Papazian DM. Potassium channel alpha and beta subunits assemble in the endoplasmic reticulum. J Biol Chem 1997; 272(5):3022-7.
40. Chan KW, Wheeler A, Csanady L. Sulfonylurea receptors type 1 and 2A randomly assemble to form heteromeric KATP channels of mixed subunit composition. J Gen Physiol 2008; 131(1):43-58.
41. Shi G, Nakahira K, Hammond S et al. Beta subunits promote K+ channel surface expression through effects early in biosynthesis. Neuron 1996; 16(4):843-52.
42. Knaus HG, Garcia-Calvo M, Kaczorowski GJ et al. Subunit composition of the high conductance calcium-activated potassium channel from smooth muscle, a representative of the mSlo and slowpoke family of potassium channels. J Biol Chem 1994; 269(6):3921-4.
43. Grahammer F, Warth R, Barhanin J et al. The small conductance K+ channel, KCNQ1: expression, function and subunit composition in murine trachea. J Biol Chem 2001; 276(45):42268-75.
44. Parcej DN, Scott VE, Dolly JO. Oligomeric properties of alpha-dendrotoxin-sensitive potassium ion channels purified from bovine brain. Biochemistry 1992; 31(45):11084-8.

45. Larsson HP, Baker OS, Dhillon DS et al. Transmembrane movement of the shaker K+ channel S4. Neuron 1996; 16(2):387-97.
46. Mannuzzu LM, Moronne MM, Isacoff EY. Direct physical measure of conformational rearrangement underlying potassium channel gating. Science 1996; 271(5246):213-6.
47. Noda M, Shimizu S, Tanabe T et al. Primary structure of Electrophorus electricus sodium channel deduced from cDNA sequence. Nature 1984; 312(5990):121-7.
48. Greenblatt RE, Blatt Y, Montal M. The structure of the voltage-sensitive sodium channel. Inferences derived from computer-aided analysis of the Electrophorus electricus channel primary structure. FEBS Lett 1985; 193(2):125-34.
49. Guy HR, Seetharamulu P. Molecular model of the action potential sodium channel. Proc Natl Acad Sci USA 1986; 83(2):508-12.
50. Bezanilla F, Perozo E, Stefani E. Gating of Shaker K+ channels: II. The components of gating currents and a model of channel activation. Biophys J 1994; 66(4):1011-21.
51. Sigg D, Stefani E, Bezanilla F. Gating current noise produced by elementary transitions in Shaker potassium channels. Science 1994; 264(5158):578-82.
52. Aggarwal SK, MacKinnon R. Contribution of the S4 segment to gating charge in the Shaker K+ channel. Neuron 1996; 16(6):1169-77.
53. Schoppa NE, McCormack K, Tanouye MA et al. The size of gating charge in wild-type and mutant Shaker potassium channels. Science 1992; 255(5052):1712-5.
54. Gandhi CS, Loots E, Isacoff EY. Reconstructing voltage sensor-pore interaction from a fluorescence scan of a voltage-gated K+ channel. Neuron 2000; 27(3):585-95.
55. Jiang Y, Ruta V, Chen J et al. The principle of gating charge movement in a voltage-dependent K+ channel. Nature 2003; 423(6935):42-8.
56. Iwasaki H, Murata Y, Kim Y et al. A voltage-sensing phosphatase, Ci-VSP, which shares sequence identity with PTEN, dephosphorylates phosphatidylinositol 4,5-bisphosphate. Proc Natl Acad Sci USA 2008; 105(23):7970-5.
57. Murata Y, Okamura Y. Depolarization activates the phosphoinositide phosphatase Ci-VSP, as detected in Xenopus oocytes coexpressing sensors of PIP2. J Physiol 2007; 583(Pt 3):875-89.
58. Murata Y, Iwasaki H, Sasaki M et al. Phosphoinositide phosphatase activity coupled to an intrinsic voltage sensor. Nature 2005; 435(7046):1239-43.
59. Ramsey IS, Moran MM, Chong JA et al. A voltage-gated proton-selective channel lacking the pore domain. Nature 2006; 440(7088):1213-6.
60. Sasaki M, Takagi M, Okamura Y. A voltage sensor-domain protein is a voltage-gated proton channel. Science 2006; 312(5773):589-92.
61. Posson DJ, Ge P, Miller C et al. Small vertical movement of a K+ channel voltage sensor measured with luminescence energy transfer. Nature 2005; 436(7052):848-51.
62. Pathak M, Yarov-Yarovoy V, Agarwal G et al. Closing in on the resting state of the Shaker K(+) channel. Neuron 2007; 56(1):124-40.
63. Treptow W, Marrink SJ, Tarek M. Gating motions in voltage-gated potassium channels revealed by coarse-grained molecular dynamics simulations. J Phys Chem B 2008; 112(11):3277-82.
64. Treptow W, Tarek M, Klein ML. Initial response of the potassium channel voltage sensor to a transmembrane potential. J Am Chem Soc 2009; 131(6):2107-9.
65. Bezanilla F. How membrane proteins sense voltage. Nat Rev Mol Cell Biol 2008; 9(4):323-32.
66. Tombola F, Pathak MM, Isacoff EY. How does voltage open an ion channel? Annu Rev Cell Dev Biol 2006; 22:23-52.
67. Schwappach B, Zerangue N, Jan YN et al. Molecular basis for K(ATP) assembly: transmembrane interactions mediate association of a K+ channel with an ABC transporter. Neuron 2000; 26(1):155-67.
68. Flagg TP, Kurata HT, Masia R et al. Differential structure of atrial and ventricular KATP: atrial KATP channels require SUR1. Circ Res 2008; 103(12):1458-65.
69. Babenko AP, Bryan J. Sur domains that associate with and gate KATP pores define a novel gatekeeper. J Biol Chem 2003; 278(43):41577-80.
70. Chan KW, Zhang H, Logothetis DE. N-terminal transmembrane domain of the SUR controls trafficking and gating of Kir6 channel subunits. EMBO J 2003; 22(15):3833-43.
71. Ueda K, Matsuo M, Tanabe K et al. Comparative aspects of the function and mechanism of SUR1 and MDR1 proteins. Biochim Biophys Acta 1999; 1461(2):305-13.
72. Ueda K, Inagaki N, Seino S. MgADP antagonism to Mg2+-independent ATP binding of the sulfonylurea receptor SUR1. J Biol Chem 1997; 272(37):22983-6.
73. Eliasson L, Renstrom E, Ammala C et al. PKC-dependent stimulation of exocytosis by sulfonylureas in pancreatic beta cells. Science 1996; 271(5250):813-5.
74. Ashcroft FM, Harrison DE, Ashcroft SJ. Glucose induces closure of single potassium channels in isolated rat pancreatic beta-cells. Nature 1984; 312(5993):446-8.

75. Alekseev AE, Hodgson DM, Karger AB et al. ATP-sensitive K+ channel channel/enzyme multimer: metabolic gating in the heart. J Mol Cell Cardiol 2005; 38(6):895-905.
76. Zingman LV, Hodgson DM, Bast PH et al. Kir6.2 is required for adaptation to stress. Proc Natl Acad Sci USA 2002; 99(20):13278-83.
77. Zingman LV, Alekseev AE, Hodgson-Zingman DM et al. ATP-sensitive potassium channels: metabolic sensing and cardioprotection. J Appl Physiol 2007; 103(5):1888-93.
78. Fan Z, Neff RA. Susceptibility of ATP-sensitive K+ channels to cell stress through mediation of phosphoinositides as examined by photoirradiation. J Physiol 2000; 529(Pt 3):707-21.
79. Tucker SJ, Gribble FM, Zhao C et al. Truncation of Kir6.2 produces ATP-sensitive K+ channels in the absence of the sulphonylurea receptor. Nature 1997; 387(6629):179-83.
80. Zerangue N, Schwappach B, Jan YN et al. A new ER trafficking signal regulates the subunit stoichiometry of plasma membrane K(ATP) channels. Neuron 1999; 22(3):537-48.
81. Koster J, Kurata HT, Enkvetchakul D et al. DEND mutation in Kir6.2 (KCNJ11) reveals a flexible N-terminal region critical for ATP-sensing of the KATP channel. Biophysical Journal 2008; 95(10):4689-97.
82. Mlynarski W, Tarasov AI, Gach A et al. Sulfonylurea improves CNS function in a case of intermediate DEND syndrome caused by a mutation in KCNJ11. Nat Clin Pract Neurol 2007; 3(11):640-5.
83. Shimomura K, Horster F, de Wet H et al. A novel mutation causing DEND syndrome: a treatable channelopathy of pancreas and brain. Neurology 2007; 69(13):1342-9.
84. Drain P, Li L, Wang J. KATP channel inhibition by ATP requires distinct functional domains of the cytoplasmic C terminus of the pore-forming subunit. Proc Natl Acad Sci USA 1998; 95(23):13953-8.
85. Antcliff JF, Haider S, Proks P et al. Functional analysis of a structural model of the ATP-binding site of the KATP channel Kir6.2 subunit. EMBO J 2005; 24(2):229-39.
86. Enkvetchakul D, Nichols CG. Gating mechanism of KATP channels: function fits form. J Gen Physiol 2003; 122(5):471-80.
87. Hollenstein K, Frei DC, Locher KP. Structure of an ABC transporter in complex with its binding protein. Nature 2007; 446(7132):213-6.
88. Dawson RJ, Locher KP. Structure of the multidrug ABC transporter Sav1866 from Staphylococcus aureus in complex with AMP-PNP. FEBS Lett 2007; 581(5):935-8.
89. Locher KP, Lee AT, Rees DC. The E. coli BtuCD structure: a framework for ABC transporter architecture and mechanism. Science 2002; 296(5570):1091-8.
90. Nakahira K, Shi G, Rhodes KJ et al. Selective interaction of voltage-gated K+ channel beta-subunits with alpha-subunits. J Biol Chem 1996; 271(12):7084-9.
91. Rhodes KJ, Keilbaugh SA, Barrezueta NX et al. Association and colocalization of K+ channel alpha- and beta-subunit polypeptides in rat brain. J Neurosci 1995; 15(7 Pt 2):5360-71.
92. Weng J, Cao Y, Moss N et al. Modulation of voltage-dependent Shaker family potassium channels by an aldo-keto reductase. J Biol Chem 2006; 281(22):15194-200.
93. Pan Y, Weng J, Cao Y et al. Functional coupling between the Kv1.1 channel and aldoketoreductase Kvbeta1. J Biol Chem 2008; 283(13):8634-42.
94. Gulbis JM, Mann S, MacKinnon R. Structure of a voltage-dependent K+ channel beta subunit. Cell 1999; 97(7):943-52.
95. Rettig J, Heinemann SH, Wunder F et al. Inactivation properties of voltage-gated K+ channels altered by presence of beta-subunit. Nature 1994; 369(6478):289-94.
96. Heinemann S, Rettig J, Scott V et al. The inactivation behaviour of voltage-gated K-channels may be determined by association of alpha- and beta-subunits. J Physiol Paris 1994; 88(3):173-80.
97. McCormack K, Connor JX, Zhou L et al. Genetic analysis of the mammalian K+ channel beta subunit Kvbeta 2 (Kcnab2). J Biol Chem 2002; 277(15):13219-28.
98. Smart SL, Lopantsev V, Zhang CL et al. Deletion of the K(V)1.1 potassium channel causes epilepsy in mice. Neuron 1998; 20(4):809-19.
99. Connor JX, McCormack K, Pletsch A et al. Genetic modifiers of the Kv beta2-null phenotype in mice. Genes Brain Behav 2005; 4(2):77-88.
100. McCormack T, McCormack K, Nadal MS et al. The effects of Shaker beta-subunits on the human lymphocyte K+ channel Kv1.3. J Biol Chem 1999; 274(29):20123-6.

CHAPTER 9

IMPLICATIONS OF 3D DOMAIN SWAPPING FOR PROTEIN FOLDING, MISFOLDING AND FUNCTION

Frederic Rousseau,[1] Joost Schymkowitz[1] and Laura S. Itzhaki*[,2]
[1]*VIB Switch Laboratory, Department of Molecular Cell Biology, Katholieke Universiteit Leuven, Leuven, Belgium;*
[2]*University of Cambridge, Department of Chemistry, Cambridge, UK*
**Corresponding Author: Laura S. Itzhaki—Email: lsi10@cam.ac.uk*

Abstract: Three-dimensional domain swapping is the process by which two identical protein chains exchange a part of their structure to form an intertwined dimer or higher-order oligomer. The phenomenon has been observed in the crystal structures of a range of different proteins. In this chapter we review the experiments that have been performed in order to understand the sequence and structural determinants of domain-swapping and these show how the general principles obtained can be used to engineer proteins to domain swap. We discuss the role of domain swapping in regulating protein function and as one possible mechanism of protein misfolding that can lead to aggregation and disease. We also review a number of interesting pathways of macromolecular assembly involving β-strand insertion or complementation that are related to the domain-swapping phenomenon.

INTRODUCTION

Three-dimensional domain swapping (referred to subsequently as "domain swapping") is the process by which two identical protein chains exchange a part of their structure to form an intertwined dimer or higher-order oligomer. The phenomenon was first proposed in the 1960s to explain the behaviour of RNase A[1] dimer and somewhat later also tryptophan synthetase[2] and tryptophanase[3] but the first crystal structures of domain-swapped proteins only emerged in the 1980s.[4-8] The terminology that is currently used was introduced in 1994 by Eisenberg and colleagues who also put forward a mechanistic framework within which to understand how and why domain swapping occurs.[9-11] In this chapter we will:

Protein Dimerization and Oligomerization in Biology, edited by Jacqueline M. Matthews.

(1) highlight recently determined structures that suggest roles for domain swapping in regulating protein function, (2) discuss quantitative studies of the energetics and kinetics of the domain-swapping process and (3) review the evidence for domain swapping as a mechanism of protein misfolding leading to aggregation and disease.

DOMAIN SWAPPING TERMINOLOGY

The structure of the subunits within the domain-swapped oligomer is identical to that of the monomer with the exception of the region that connects the exchanging domain with the rest of the protein (Figs. 1, 2A). In most cases, this so-called "hinge loop" region folds back on itself to form the monomer and adopts an extended conformation in the domain-swapped dimer. Although the process is known as "domain" swapping, proteins are often found to swap only a single secondary structure element such as a β-strand or α-helix rather than a whole domain of structure. The swapped structure can be located in any part of the polypeptide sequence although it is generally at the N- or C-terminus. In some cases, approximately half of the molecule is swapped and it is therefore difficult to define which half constitutes the swapped domain. The interactions made between the swapped domain and the rest of the protein are the same in the oligomer as in the monomer but they are formed in an inter- rather than a intramolecular fashion. These intermolecular interactions comprise the "primary" interface. Since the subunits are often close to each other in the domain-swapped oligomer, an intermolecular new interface may be created that is not present in the monomeric form and this is known as the "secondary" interface. Domain swapping can potentially occur in a reciprocal manner to form a dimer, in a cyclical manner to form a trimer, tetramer etc., or in an open-ended manner to form an oligomer leaving uncomplemented ends available to assemble further. In order to be classed as a domain-swapping protein, both monomer and domain-swapped forms need to have been observed. In some cases, however, there is a structure of the domain-swapped form of a protein but no structure of the closed monomer and the protein is therefore considered to be a 'candidate' for domain swapping. In other cases, the protein has a homolog that is a closed monomer; these oligomers are classed as "quasi domain swapped".

DOMAIN-SWAPPED STRUCTURES AND REGULATION OF PROTEIN FUNCTION

Many of the proteins and protein domains that are commonly used as model systems for studying protein folding and molecular recognition have been crystallized as domain-swapped forms in addition to the monomeric forms, most notably SH2[12] and SH3[13]domains, staphylococcal nuclease, chymotrypsin inhibitor 2[14] (all domain-swapped dimers) and barnase (a domain-swapped trimer).[15] These examples suggest that, although domain swapping is relatively rare (there are less than 60 structures of domain-swapped proteins to date), many proteins have regions with features suggesting they could act as hinge loops and it may therefore be possible to induce many proteins to domain swap simply by making a few amino acid substitutions. The potential ease with which a domain-swapped species may become stabilized suggests that domain swapping could be a mechanism for the evolution of larger, complex folds from smaller, simpler ones via a domain-swapped intermediate followed by a gene duplication or fusion process.[11] There is a subset of proteins for which evidence suggests

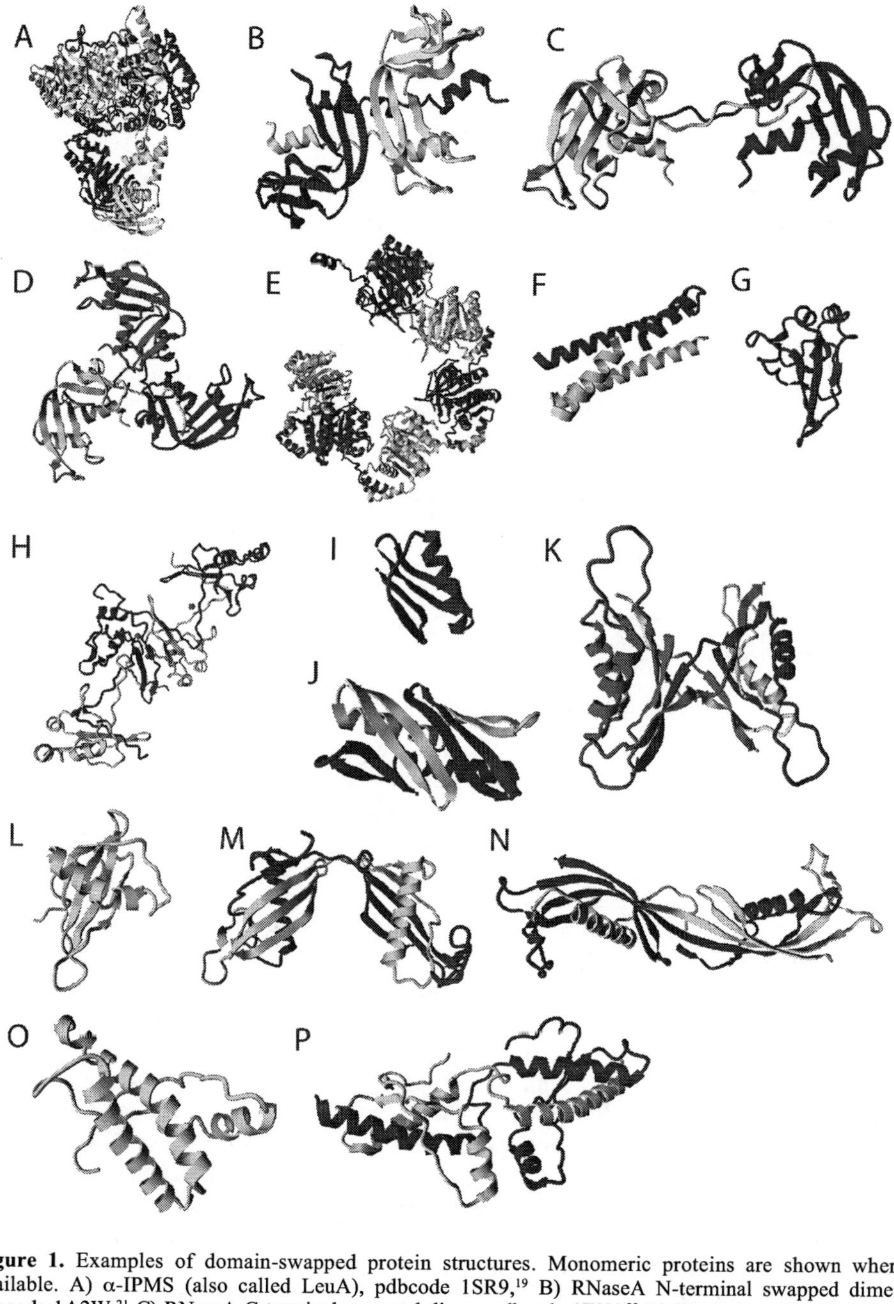

Figure 1. Examples of domain-swapped protein structures. Monomeric proteins are shown where available. A) α-IPMS (also called LeuA), pdbcode 1SR9,[19] B) RNaseA N-terminal swapped dimer, pdbcode 1A2W,[21] C) RNaseA C-terminal swapped dimer, pdbcode 1F0V,[22] D) RNaseA cyclic C-terminal trimer, pdbcode 1JS0,[70] E) T7 helicase filament structure, pdbcode 1CR4,[71] F) Designed 3-helix bundle, pdbcode 1G6U,[28] G) suc1 monomer[32] and H) domain-swapped dimer, pdbcode 1SCE,[33] I) wild-type GB1 monomer, pdbcode 1PGP,[72] J) mutant GB1 domain-swapped dimer, pdbcode 1Q10[37] and K) mutant GB1 doubly domain-swapped tetramer, pdbcode 1MPE,[38] L) Cystatin monomer, pdbcode 1GD3,[73] M) Cystatin domain-swapped dimer, pdbcode 1G96,[40] N) Cystatin amyloid-like domain-swapped dimer, pdbcode 1tij,[74] O) Human prion protein monomer, pdbcode 1QM0[51] and P) domain-swapped dimer, pdbcode 1I4M.[39]

that domain swapping regulates function and we shall focus on examples of these proteins here. For other proteins, it remains to be seen whether or not domain swapping is simply an artefact of the high protein concentrations associated with the crystallization process.

SCAN Domain of HIV-1 Capsid C-Terminal Domain

Formation of the immature retrovirus particle is directed by interactions of the protein Gag with itself, with an arrangement of several thousand copies of the protein. After particle formation, maturation occurs by proteolytic processing of Gag to release the proteins found in the infectious virus. The characteristic conical core structure is formed by the proteolytically-released capsid protein (CA) around the viral genome. The C-terminal domain of CA (CA-CTD) contains the most highly conserved sequence within Gag, a 20-residue sequence known as the major homology region (MHR). Viral assembly and/ or infectivity are sensitive to mutations or deletions in this region. Until recently it was thought that CA-CTD functions as a dimerization domain although the dimer interface does not include the MHR and other CA's do not dimerize in this way and it was therefore unclear how this interaction could account for the critical role played by CA-CTD in viral assembly. Then two structures emerged: a SCAN domain which is structurally and evolutionary related to the retroviral CA-CTD; and, the other the HIV-1 CA-CTD.[16,17] Domain swapping was induced in the latter by deletion of a single amino acid in the hinge loop region. These structures revealed that the swapped element encompasses the MHR. A functional role of domain swapping would therefore rationalize the sequence conservation of the MHR sequence and would also explain a number of other features of viral assembly. However, whether or not the CA is domain swapped in the mature viral core will await the determination of higher resolution structures of this particle.

Forkhead Domain of FOXP2

FOXP2 is a member of newly defined subfamily of the forkhead box (FOX) transcription factors and many disease-causing mutations are found in the forkhead domains of these proteins. The crystal structure of FOXP2 bound to DNA revealed that it forms a domain-swapped dimer.[18] In classical FOX proteins, a proline is present in a short turn connecting two helices. Replacement of this proline by alanine in FOXP2 results in the formation of a single long helix, thereby rigidifying the polypeptide chain in this region (see also later section on the energetic determinants of domain swapping) and causing it to swap into another molecule. Mutation of the alanine in FOXP2 to proline was found to prevent domain swapping. Alanine is present at this position in all FOXP members, suggesting that domain swapping is a conserved feature and may therefore be required for function. Modelling predicts that, due to the arrangement of the DNA binding surfaces, domain-swapped FOXP dimers can only bind cognate DNA sites that are well separated from each other, or located on different DNA strands, which suggests that these proteins loop DNA or mediate interchomosomal associations. Moreover, disease-associated mutations map to the domain-swapped dimer interface.

α-IPMS

α-IPMS from *mycobacterium tuberculosis* is an α-isopropylmalate synthase which catalyses the first committed step of the essential leucine biosynthetic pathway in this organism. The structure of the protein reveals that the 70 kDa momomer folds into two major domains separated by two much smaller linker domains that are joined by a flexible hinge[19] (Fig. 1A). The dimer is created by swapping of the major domains between two monomers. The dimer interface is extensive and spans both major domains but not the middle linker domains. 7650 Å of the monomer surface area is buried upon domain swapping, representing a quarter of the total monomer surface. The linker region appears flexible, giving rise to different relative orientations of the N- and C-terminal domains in the two monomers. The catalytic site is located within the N-terminal domain (a TIM barrel). The C-terminal domain is proposed to be the regulatory domain that is required for leucine feedback inhibition. Leucine binds to this domain, with the binding site lying between two helices, one from each of the two monomers. Other enzymes that are subject to end-product regulation bind their product inhibitors on domains remote from the catalytic domain and undergo major conformational changes on product binding; however, this is not the case for α-IPMS. It is thought instead that binding of leucine to the regulatory domain at the dimer interface is communicated to the distant catalytic domain by means of the flexible domain linker.

RNase A

RNase A is probably the best studied domain-swapped protein to date. RNase A is particular interesting because of its ability to swap two different domains, termed "double domain swapping". RNase A or bovine seminal RNase can swap either an N-terminal (Fig. 1B) or C-terminal region (Fig. 1C and D), with more stringent unfolding conditions favouring the C-terminal swapping.[20-24] A trimeric form has also been crystallized in which both N- and C-terminal regions are swapped.[25] This feature potentially allows the protein to adopt a variety of differently assembled oligomeric states, including branched structures. Domain swapping in bovine seminal RNase may also be important for function. Domain-swapped dimerisation gives rise to two composite active sites with residues from both subunits contributing to each one and there is cooperativity between the sites.[26] Remarkably, the domain-swapped dimer displays selective toxicity for tumour cells whereas the monomer does not.[27]

ENERGETIC DETERMINANTS OF DOMAIN SWAPPING

Hinge Loop Length

The major determinant of domain swapping is the hinge loop, which is is the only region of the protein that adopts a different conformation in monomeric and domain-swapped forms (excluding those dimers in which a secondary interface forms.) The easiest way to increase the domain-swapping propensity of a protein is to shorten the hinge loop, thereby making it harder for the polypeptide chain to fold back on itself. Loop deletion is seen in a number of natural proteins and this strategy has been successfully applied to a number of proteins to induce them to domain swap (Fig. 2B). A particularly nice example of this strategy comes from the design of two three-helix bundles of different topologies. Loop deletion in one leads to the formation of a reciprocal domain-swapped dimer whereas loop deletion in the other leads to the formation of fibrils by open-ended domain swapping (Fig. 1F and 2C).[28]

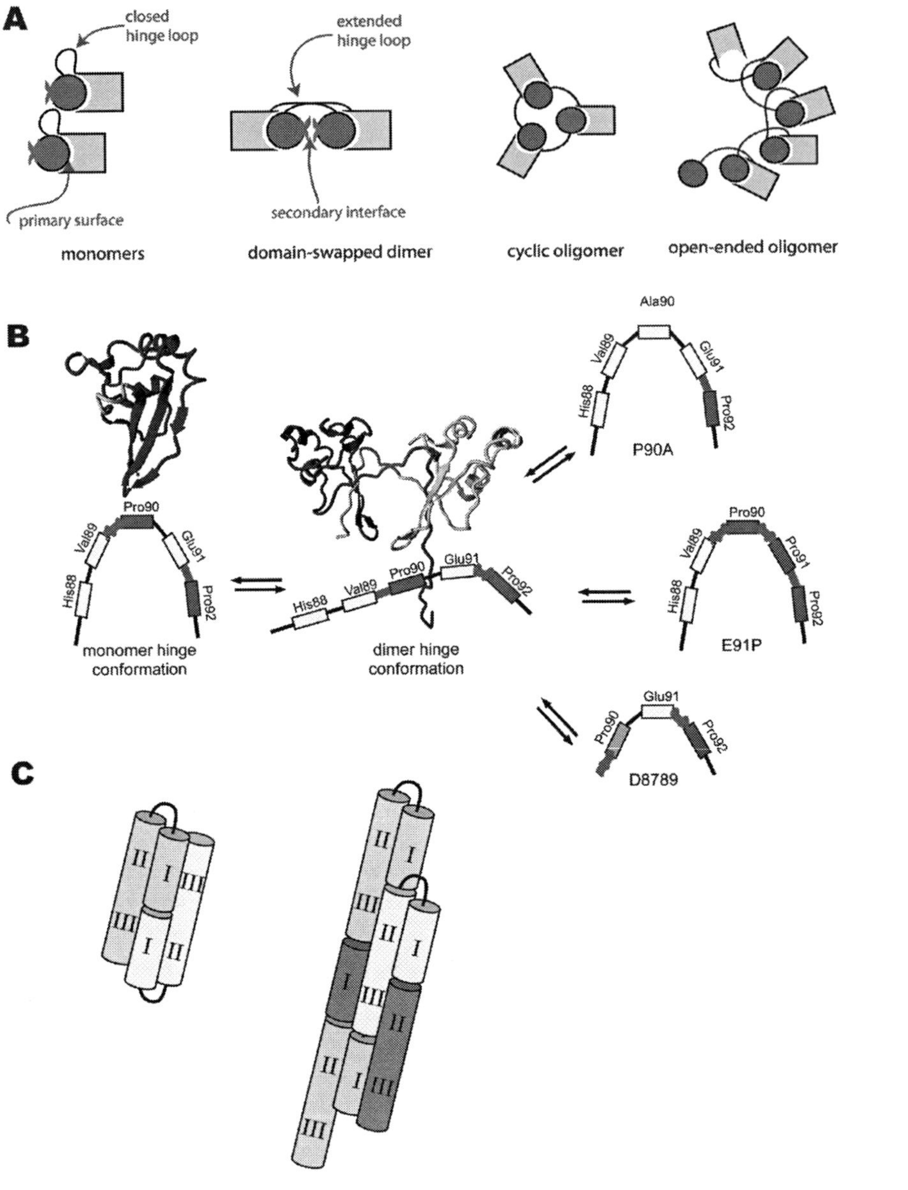

Figure 2. Schematic representation of the key features of domain-swapped structures and of domain-swapping mechanisms. A) Scheme of domain swapping showing the hinge loop, primary and secondary interfaces highlighted in red and the formation of cyclical and open-ended domain swapping in addition to reciprocal domain swapping. B) Scheme of domain swapping in suc1 showing the consequences of strain in the hinge loop imposed by proline residues. Strain is indicated by the zig-zag lines. Mutating the proline residues in the hinge reduces the strain and alters the monomer-dimer equilibrium. Shortening the hinge loop is another method of shifting the equilibrium by imposing strain on the closed monomer hinge conformation and thereby favouring the dimer conformation. C) Schematic of the design of two different three-helix bundle topologies, one of which produces a reciprocal domain-swapped dimer and the other an open-ended domain-swapped oligomer. A color version of this figure is available online at www.landesbioscience.com/curie.

Hinge Loop Flexibility

A second strategy for engineering domain-swapped dimers has been to introduce greater flexibility into the hinge loop, either by lengthening it or by mutation. A polyglutamine insertion into the long active site loop of chymotrypsin inhibitor 2 resulted in a domain-swapped dimer and higher-order oligomers also.[29] The extent of higher-order oligomer formation increased with loop insertions consisting of repeats of glutamine, alanine and glycine, in that order. Likewise for the protein suc1 (Fig. 1G,H), the wild-type protein forms only monomers and domain-swapped dimers at low millimolar concentrations but substitution of both proline residues in the hinge loop for alanines resulted in the formation of trimers and higher-order oligomers and this effect was enhanced when glycine residues were substituted for the prolines.[30]

Proline Residues in Hinge Loops

The proline residues found in the hinge loop of the protein suc1 appear to play a pivotal role in modulating its domain-swapping propensity. Indeed, Bergdoll and coworkers had observed previously that proline residues are found in the hinge loops of a number of proteins that domain swap.[31] The results for suc1 indicated that proline residues create strain in the hinge loop (Fig. 2B). Suc1 is a 113-residue protein comprising a 4-stranded β-sheet that packs against two α-helices.[32,33] Domain swapping in suc1 occurs by the exchange of a C-terminal sequence that comprises a central strand of the β-sheet (Fig. 1G,H). Mutation of one or other proline residue in the hinge loop to alanine greatly shifts the monomer-dimer equilibrium, changing the dissociation constant by several orders of magnitude, whereas mutation of other residues in the hinge loop to alanines has very little effect.[30] Mutation of the first proline to alanine stabilizes the monomer form, whereas mutation of the second proline to alanine stabilizes the dimer form. The hinge loop is in a strained conformation in both monomer and dimer forms of the protein, more so in the monomer (Fig. 2B). The strain appears to make the hinge loop act as a loaded molecular spring that releases tension present in the monomer by adopting an alternative conformation in the dimer.[30] The proline residues do not appear to destabilize the hinge loop by making unfavourable interactions with other residues but rather by imposing strain on the backbone. Similar behaviour was observed for protein L.[34] Mutations were designed to destabilize the hinge loop conformation of the monomer of protein L by forcing residues into forbidden regions of the Ramachandran plot, thereby inducing an alternative, domain-swapped form.[34]

Mutation of Residues Distant from the Hinge Loop

The domain-swapping potential of the protein CD2 could be altered by making mutations in the secondary interface, the new intermolecular interface that is created in the domain-swapped form. Somewhat surprisingly, however, it has been found for domain-swapped proteins that lack a secondary interface (i.e., proteins for which the interactions made in the domain-swapped structure are identical to those in the monomer with the exception of the hinge loop), that mutations outside of the hinge loop can also alter the domain-swapping propensity. These observations suggest signalling, i.e., that sites distant from the hinge loop can sense its conformation. In the case of suc1, mutation of residues in the β-sheet that were distant from the hinge loop was found to alter the equilibrium between monomer and dimer. The effects were not as large as for mutations in the hinge loop, but they were nevertheless significant. Likewise, a site in the β-sheet distant from the hinge loop was found to bind phosphorylated ligands with different affinities in the monomer and the dimer forms. The crystal structures of the monomer and domain-swapped dimer are highly superimposable outside of the hinge loop region and therefore they provide no explanation for these effects. The different energetics of the monomer and dimer forms must instead result from a difference in the dynamic properties of the conformational ensemble of the native states that is present in solution. It was proposed that the strain, imposed on the hinge loop by the proline residues, is accommodated differently in the two forms. Consistent with this idea, removing the strain by mutating the proline residues to alanine reduces the difference in binding affinities of the two forms. NMR and molecular dynamics simulations of cks1, the human homolog of suc1, further support the model.[35] The experiments revealed conformational heterogeneity in the β-sheet of the monomer; similar behaviour has also been noted for suc1.[36] The region of greatest conformational variability mapped to the domain-swapping β-strand. Upon binding of a conserved ligand, cdk2, to one face of the β-sheet containing the hinge loop region, the heterogeneity was frozen out and this complexed form of cks1-bound phosphorylated ligand on the opposite face of the β-sheet with higher affinity than the free form. The observation is consistent with the signal transduction in the β-sheet observed in the domain-swapping experiments. It was therefore proposed that conformational heterogeneity is required for cks function and that a side effect of this requirement is that the protein domain has the ability to domain swap.

An example of mutations changing a protein's tertiary and quaternary structure in a wholly unanticipated way is provided by the immunoglobulin-binding domain B1 of streptococcal protein G (GB1). GB1 is a 56-residue protein, comprising a 4-stranded β-sheet packing against a single α-helix (Fig. 1I). In the course of screening hydrophobic core mutants for changes in stability some striking results were obtained. A domain-swapped dimer was formed that had the same structural elements as the wild-type monomer but the intertwined nature of the swapping, involving half of the secondary structural elements, was such as to create a significant secondary interface[37] (Fig. 1J). A further mutation led to an even more startling result. A double domain-swapped tetramer was formed in which the structural elements were very different from the wild-type monomer[38] (Fig. 1K). Closer inspection of the structures led the authors to conclude that destabilisation or opening up of the hydrophobic core upon mutation was compensated by extending the core via oligomerisation. Consistent with the idea that the mutations significantly destabilize the monomer structure, the monomeric form of this mutant variant is only partially folded with characteristics of a molten-globule state (i.e., secondary structural elements present but not fixed by tertiary contacts).[38]

KINETIC MECHANISMS OF DOMAIN SWAPPING

For domain swapping to occur, it is necessary for the protein to unfold in order for it to release the domain and then reassemble in an intertwined form. The mechanism of how a protein domain swaps and of how monomer and domain-swapped dimer forms interconvert, is therefore essentially a protein folding problem. For many proteins, interconversion is very slow under physiological conditions, consistent with idea of some degree of unfolding being required. The process is speeded up under conditions that favour unfolding. Eisenberg and coworkers proposed that the transition state for the interconversion reaction is an "open" form of the structure, in which the interactions between the exchanging domain and the rest of the protein are broken but the native folds within these two parts are retained. Such a scenario is most likely for proteins that swap a true domain capable of folding independently of the rest of the structure which can then assemble. However, many proteins swap only a small number of secondary structural elements that do not constitute a separate domain. In such cases, it is more likely that folding and association occur in a more concomitant fashion. Suc1 provides an extreme example of this sort of domain swapping, whereby the exchanging "domain" is a single strand of a 4-stranded β-sheet. Although this strand is at the very C-terminus of the polypeptide chain it is a central strand in the β-sheet and therefore integral to its stability. Protein engineering analysis and molecular dynamics simulations of the folding mechanisms of both monomeric and domain-swapped dimeric suc1 revealed that the interactions between this strand and the rest of the protein are formed very early in the folding pathway (and, conversely, very late in the unfolding reaction). Therefore, the entire protein must unfold substantially in order to interconvert between monomer and dimer. Studies of the folding of the N-terminal, Ig domain of CD2, engineered to domain swap by deletion of residues in the hinge loop, indicate similar behaviour. However, in this case folding proceeds via intermediate in which one-half of the domain-swapped dimer is folded and the other half is not.

DOMAIN SWAPPING AND REFOLDING

The folding reactions of the monomer and the domain-swapped dimer of suc1 are initiated at the same site (or folding nucleus), the only difference being that some of the key interactions of this nucleus are formed intra- vs. inter-molecularly in the two reactions. Consequently, there is ambiguity in the refolding process, which can lead to two different products. The competition between the collapse of a chain on itself to form a monomer and the probability of interacting with another chain before the key interactions are formed is affected by loop lengthening. Loop lengthening slows down the rate of folding of the monomer because of the higher entropic cost of fixing a longer loop than a shorter one and this effect indirectly favours folding to the dimer because it increases the likelihood of association with another chain. An unanticipated result for suc1 was that, whereas mutations in suc1 can alter the domain-swapping propensity at equilibrium by many orders of magnitude, the product of refolding of all these mutant variants is overwhelmingly the monomeric form. Thus, the domain-swapped dimer may be the more stable form for certain of the suc1 mutants but it is not kinetically accessible. Evidence suggests that the population of an intermediate state acts to buffer the monomeric folding pathway against the effect of mutations that favour the dimer at equilibrium.

Another effect of loop lengthening is to increase the tendency to aggregate. Many proteins, including CI2, U1A and maltose binding protein, undergo transient association on refolding at moderate to high concentrations and domain-swapping provides a mechanism for such a phenomenon. The formation of higher order oligomers could occur through the same native nucleus by domain swapping in an open-ended, rather than reciprocal manner. These oligomers are only transient because of the uncomplemented ends that are produced by open-ended domain swapping.

DOMAIN SWAPPING, PROTEIN MISFOLDING AND AGGREGATION

With the discovery of domain swapping, Eisenberg realized that the phenomenon could be a way for a protein to form aggregates such as those found in disease. Subsequently, two disease-associated proteins, prion protein and cystatin, were crystallized in domain-swapped forms[39,40](Fig. 1M-P). There is now increasing evidence that domain swapping is involved in the assembly of some proteins into ordered aggregates. Of the various models for structures of amyloid fibrils that are consistent with its characteristic cross-β diffraction pattern, domain swapping has been implicated in three different ways (reviewed in Nelson and Eisenberg (2006)[41]). An important common feature of these models, distinct from those models in which the whole protein is envisaged to convert from its native structure to a β-sheet conformation, is that the subunits of the fibril retain much of the native structure of the monomeric form of the protein. This aspect of the domain-swapping models is particularly appealing in view of the fact that amyloid formation in vivo takes place under physiological conditions (albeit facilitated by the likely destabilizing nature of disease-associated variants) rather than the harshly denaturing conditions commonly used in vitro to induce fibril formation. Moreover, several disease-associated proteins, such as the prion protein, β2-microglobulin and cystatin, have disulphide bonds and therefore amyloid formation involving large-scale unfolding of the structure would not be possible under nonreducing conditions. According to the cross-β spine model, a small segment of the polypeptide chain has the tendency to stack into a β-sheet.[42] The segment may be located at the end of a folded domain or between folded domains and these domains are natively structured in the fibril. Such a model has been proposed for fibrils of β2-microglobulin, in which the β-sheet stacking region is at the N-terminus and the rest of the protein remains natively folded.[43] If the stacking region is in the middle of the polypeptide chains in between sub-domains, then these may still be able to reform the native structure by complementing each other via domain swapping. Such a structure may be present in fibrils from a designed variant of RNase A containing a polyglutamine insertion (Q10) in the hinge loop.[44] Evidence for the model comes from the observation that fibrils made by mixing two Q10 variants, each of which contains a mutated catalytic residue on either the core domain or the swapped domain, have RNase activity.

In the second model, fibril formation occurs by domain swapping without formation of a cross-β spine, as has been proposed for fibrils of cystatin and β2-microglobulin.[40,45,46] Both proteins have β-sheet structures and reciprocally domain-swapped dimers or oligomers could pack against each other to form a fibril. The potential for domain-swapped dimers to produce high-order assemblies in this way was pointed out by Murray et al for the β-sheet N-terminal domain of the protein CD2.[47] Likewise, a crystal-wide β-sheet structure was also observed for the llama VHH-R9 domain built up of stacked domain-swapped dimers.[48] According to the third model, domain swapping could occur in an open-ended or runaway manner to produce high-order oligomers, as has been seen for amyloid fibrils of T7 endonuclease and for trpR, the latter forming a double domain-swapped crystalline array.[49,50] Some more detailed examples, including possible pathways by which fibrils are formed from native monomers, are given below.

Prion Protein

The structure of the human prion protein has been solved as a monomer and also as a domain-swapped dimer.[39,51](Fig. 1O,P). A disulphide bond links the swapped domain and the remaining protein and therefore monomer-dimer conversion requires reduction of the disulphide bond and then reoxidation. In a set of elegant experiments, Lee and Eisenberg took note of this disulphide bond linkage and were able to develop the first time an in vitro redox method of converting recombinant PrPc (its normal cellular form) to an infectious form they term PrP^{RDX}, which has similar properties to PrP^{Sc}. PrP^{RDX} could seed the conversion of PrP^{C} to PrP^{RDX}, and had double the β-sheet content of PrP^{C}.[52] They proposed a model for the conversion process involving runaway domain swapping in which the hinge loops of successive subunits form a continuous β-sheet at the centre of the fibril. The domain-swapped PrP globular domains sit on the outside of the β-sheet. The authors highlight the fact that a domain-swapping mechanism for the recruitment of normal protein into the fibrils is able to explain numerous features of prion diseases. In particular, the stability of both forms of the prion protein can be rationalized by the high energy barrier required to break both noncovalent and covalent (disulphide bonds) interactions. Also, self-propagation is able to occur because domain-swapping acts as a templating mechanism.

GB1

Strikingly, those mutant variants of GB1 that can exist as domain-swapped dimers (described earlier in this chapter) form fibrils readily, whereas those that fold only into the monomer form do not.[53] A model for the fibrils was constructed involving packing of domain-swapped dimers against each other via an edge β-strand. This arrangement results in a contiguous β-sheet in which the individual β-strands run perpendicular to the long axis consistent with the diffraction pattern of the fibres. Residue Ala34 that switches the protein from a monomer to a domain-swapped dimer was proposed to act as a gatekeeper residue that ensures the protein folds efficiently in the momoeric species and does not misfold into alternative forms. When replaced by larger hydrophobic residues that force the core to expand and be destabilized, alternative conformations become accessible.

Suc1

Early on in the domain-swapping studies of suc1, a correlation was observed between domain swapping and aggregation propensities of mutant variants. Electron microscopy revealed that suc1 aggregates have the appearance of a string-of-pearls of dimensions consistent with a runaway domain-swapped arrangement of the molecules. Further experiments verified that the protein in the aggregates was indeed domain-swapped and native-like in structure. These aggregates could be formed by incubating protein at millimolar concentrations and room temperature or by refolding acid-denatured protein at concentrations in the high micromolar range. Interestingly, it was found that mutants with very low domain-swapping propensities and showing negligible aggregation could be induced to aggregate by refolding in the presence of wild type or domain-swapping prone mutants. The domain-swapped suc1 aggregates were cytotoxic whereas amorphous aggregates, formed by heating the protein, were not. This result appears counterintuitive at first, given the native-like character of the domain-swapped aggregates. However, runaway domain swapping leaves uncomplemented ends and these unfolded elements are likely to be sticky and could potentially interact with various cellular components impairing their function. The observed toxicity is reminiscent of numerous other studies showing that nonfibrillar and prefibrillar assemblies are toxic but the fibril end-products are not. It was proposed that domain swapping could be one mechanism for the early stages of fibril formation that brings together amyloid-prone segments from different polypeptide chains, thereby facilitating further assembly into fibrils.

STRAND INSERTION AND COMPLEMENTATION: SERPINS, PILUS ASSEMBLY AND Rad51-BRCA2

We conclude this chapter by pointing out that there are other biological assemblies that are reminiscent of domain swapping. In each case, a strand is provided by another molecule of the same protein or of a different protein to complete or expand a β-sheet. Rad51 is a recombinase enzyme that forms nucleofilaments and both Rad51 and its homolog RecA have been shown to form filaments by runaway swapping of a β-strand (Fig. 1E). BRCA2, commonly mutated in breast, ovarian and other cancer types, controls the function of Rad51 in pathways for DNA repair by homologous recombination. The recently solved structure of the Rad51-BRCA2 complex reveals how BRCA2 inhibits the assembly of Rad51 nucleofilaments by mimicking the β-strand exchange process.[54]

Specific molecular chaperones help to assemble pilin subunits into adhesive pili, which are rod-like structures that allow pathogenic Gram-negative bacteria to adhere to and colonize host tissues. Pilus subunits lack the seventh β-strand (G1 strand) that would complete their Ig fold and so expose a hydrophobic groove on their surface. Crystal structures of pilus-chaperone complexes indicate that the chaperone donates a β-strand from its Ig-like domain, thereby completing the Ig fold of the pilus subunit, although interestingly the chaperone G1 strand runs parallel to the F strand of the pilus subunit rather than antiparallel in the true Ig fold.[55,56] A model has been proposed in which the chaperone carries a pilus subunit to the large pore protein (the usher) where the pilus subunit is released and becomes attached to the end of the growing pilus rod (reviewed in 57). This process occurs by an N-terminal extension of the incoming pilus subunit inserting into the end subunit of the rod, replacing the strand temporarily donated by the chaperone. The process is termed "donor strand complementation".

Serpins (serine protease inhibitors) contain an exposed and mobile reactive loop that presents a peptide sequence as a pseudosubstrate for the target proteinase. Upon docking of the serpin with the protease, a large conformational change occures in the former resulting in the insertion of the cleaved reactive loop as an additional strand into the s-stranded β-sheet of the serpin and consequent inhibition of the protease.[58] Disease-associated point mutations act by destabilising the β-sheet allowing the incorporation of the reactive loop of another serpin molecule and thereby leading to the formation of an extended polymer chain.[59,60] The Z mutation in α_1-antitrypsin, associated with liver disease, is a glutamic acid to lysine substitution at the base of the mobile loop. The Z mutation results in the accumulation of protein in inclusions in the rough endoplasmic reticulum of the liver.[61] Z α_1-antitrypsin has also been shown to form chains of polymers in vitro when incubated under physiological conditions.[59,62] By the same mechanism, inclusions of a mutant neuroserpin are associated with FENIB (familial encephalopathy with neuroserpin inclusion bodies).[63-65]Interestingly, mutations in the protein antithrombin associated with premature thrombosis, instead of causing polymerisation, result in the formation of inactive 6-stranded monomer that then binds to a normal antithrombin molecule leading to propagation of the conformational inactivation, similar to the proposed prion mechanism.[59,66-69] The process has been shown to occur both in vitro and in vivo.

CONCLUSION

The last few years have seen a significant number of papers delineating the relationship between domain swapping, misfolding and disease.[75-79] The work on simple model (usually non-disease associated) proteins has helped in understanding the more complex disease-associated proteins. In parallel, computational studies, in particular using funneled energy landscape models, have provided a conceptual framework for understanding the structural determinants of domain swapping and the mechanisms by which it occurs.[80-84] In future, we look forward to filling in further details of how domain swapping contributes to both the function and the malfunction of proteins.

REFERENCES

1. Crestfield AM, Stein WH, Moore S. On the aggregation of bovine pancreatic ribonuclease. Arch Biochem Biophys 1962; 1:217-222.
2. Jackson DA, Yanofsky C. Restoration of enzymatic activity by complementation in vitro between mutant alpha subunits of tryptophan synthetase and between mutant subunits and fragments of the alpha subunit. J Biol Chem 1969; 244:4539-4546.
3. London J, Skrzynia C, Goldberg ME. Renaturation of Escherichia coli tryptophanase after exposure to 8M urea. Evidence for the existence of nucleation centers. Eur J Biochem 1974; 47:409-415.
4. Anderson WF, Ohlendorf DH, Takeda Y et al. Structure of the cro repressor from bacteriophage lambda and its interaction with DNA. Nature 1981; 290:754-8.
5. Fita I, Rossmann MG. The NADPH binding site on beef liver catalase. Proc Natl Acad Sci 1985; 82:1604-1608.
6. Parge HE, Arvai AS, Murtari DJ et al. Human CksHs2 atomic structure: a role for its hexameric assembly in cell cycle control. Science 1993; 262:387-95.
7. Remington S, Wiegand G, Huber R. Crystallographic refinement and atomic models of two different forms of citrate synthase at 2.7 and 1.7 A resolution. J Mol Biol 1982; 158:111-52.
8. Story RM, Weber IT, Steitz TA. The structure of the E. coli recA protein monomer and polymer. Nature 1992; 355:318-25.
9. Bennett MJ, Schlunegger MP, Eisenberg D. 3D domain swapping: A mechanism for oligomer assembly. Prot Sci 1995; 4:2455-2468.
10. Bennett MJ, Choe S, Eisenberg D. Domain swapping: entangling alliances between proteins. Proc Nat Acad Sci USA 1994; 91:3127-3131.
11. Schlunegger MP, Bennett MJ, Eisenberg D. Oligomer formation by 3D domain swapping: a model for protein assembly and misassembly. Adv Protein Chem 1997; 50:61-122.
12. Schiering N, Casale E, Caccia P et al. Dimer formation through domain swapping in the crystal structure of the Grb2-SH2-Ac-pYVNV complex. Biochemistry 2000; 39:13376-13382.
13. Kishan KVR, Newcomer ME, Rhodes TH et al. Effect of pH and salt bridges on structural assembly: Molecular structures of the monomer and intertwined dimer of the Eps8 SH3 domain. Protein Science 2001; 10:1046-1055.
14. Chen YW, Stott K, Perutz MF. Crystal structure of a dimeric chymotrypsin inhibitor 2 mutant containing an inserted glutamine repeat. Proc Natl Acad Sci USA 1999; 96:1257-61.
15. Zegers I, Deswarte J, Wyns L. Trimeric domain-swapped barnase. Proc Natl Acad Sci USA 1999; 96:818-822.
16. Ivanov D, Stone JR, Maki JL et al. Mammalian SCAN domain dimer is a domain-swapped homolog of the HIV capsid C-terminal domain. Mol Cell 2005; 17:137-43.
17. Sawaya MR, Sambashivan S, Nelson R et al. Atomic structures of amyloid cross-beta spines reveal varied steric zippers. Nature 2007; 447:453-7.
18. Stroud JC, Wu Y, Bates DL et al. Structure of the forkhead domain of FOXP2 bound to DNA. Structure 2006; 14:159-66.
19. Koon N, Squire CJ, Baker EN. Crystal structure of LeuA from Mycobacterium tuberculosis, a key enzyme in leucine biosynthesis. Proc Natl Acad Sci USA 2004; 101:8295-300.
20. Mazzarella L, Capasso S, Demasi D et al. Bovine seminal ribonuclease—structure at 1.9-Angstrom resolution. Acta Crystallogr D Biol Crystallogr 1993; 49:389-402.
21. Liu Y, Hart PJ, Schlunegger MP et al. The crystal structure of a 3D domain-swapped dimer of RNase A at a 2.1- A resolution. Proc Natl Acad Sci USA 1998; 95:3437-42.

22. Liu Y, Gotte G, Libonati M et al. A domain-swapped RNase A dimer with implications for amyloid formation. Nat Struct Biol 2001; 8:211-4.
23. Esposito L, Daggett V. Insight into ribonuclease A domain swapping by molecular dynamics unfolding simulations. Biochemistry 2005; 44:3358-68.
24. Gotte G, Vottariello F, Libonati M. Thermal aggregation of ribonuclease A. A contribution to the understanding of the role of 3D domain swapping in protein aggregation. J Biol Chem 2003; 278:10763-9.
25. Liu Y, Gotte G, Libonati M et al. Structures of the two 3D domain-swapped RNase A trimers. Protein Sci 2002; 11:371-80.
26. Piccoli R, Di Donato A, D'Alessio G. Co-operativity in seminal ribonuclease function. Kinetic studies. Biochem J 1988; 253:329-36.
27. Didonato A, Cafaro V, Romeo I et al. Hints On the Evolutionary Design of a Dimeric Rnase With Special Bioactions. Protein Science 1995; 4:1470-1477.
28. Ogihara NL, Ghirlanda G, Bryson JW et al. Design of three-dimensional domain-swapped dimers and fibrous oligomers. Proc Natl Acad Sci USA 2001; 98:1404-9.
29. Gordon-Smith DJ, Carbajo RJ, Stott K et al. Solution studies of chymotrypsin inhibitor-2 glutamine insertion mutants show no interglutamine interactions. Biochemical and Biophysical Research Communications 2001; 280:855-860.
30. Rousseau F, Schymkowitz JWH, Wilkinson HR et al. Three-dimensional domain swapping in p13suc1 occurs in the unfolded state and is controlled by conserved proline residues. Procl Natl Acad Sci USA 2001; 98:5596-5601.
31. Bergdoll M, Eltis LD, Cameron AD et al. All in the family: Structural and evolutionary relationships among three modular proteins with diverse functions and variable assembly. Protein Science 1998; 7:1661-1670.
32. Endicott JA, Noble ME, Garman EF et al. The crystal structure of p13suc1, a p34cdc2-interacting cell cycle control protein. EMBO J 1995; 14:1004-1014.
33. Bourne Y, Arvai AS, Bernstein SL et al. Crystal structure of the cell cycle-regulatory protein suc1 reveals a beta-hinge conformational switch. Procl Natl Acad Sci USA 1995; 92:10232-10236.
34. O'Neill JW, Kim DE, Johnsen K et al. Single-Site Mutations Induce 3D Domain Swapping in the B1 Domain of Protein L from Peptostreptococcus magnus. Structure (Camb) 2001; 9:1017-1027.
35. Seeliger MA, Spichty M, Kelly SE et al. Role of conformational heterogeneity in domain swapping and adapter function of the Cks proteins. J Biol Chem 2005; 280:30448-59.
36. Landrieu I, Odaert B, Wieruszeski JM et al. p13SUC1 and the WW Domain of PIN1 Bind to the Same Phosphothreonine- Proline Epitope. J Biol Chem 2001; 276:1434-1438.
37. Byeon IJ, Louis JM, Gronenborn AM. A protein contortionist: core mutations of GB1 that induce dimerization and domain swapping. J Mol Biol 2003; 333:141-52.
38. Kirsten Frank M, Dyda F, Dobrodumov A et al. Core mutations switch monomeric protein GB1 into an intertwined tetramer. Nat Struct Biol 2002; 9:877-85.
39. Knaus KJ, Morillas M, Swietnicki W et al. Crystal structure of the human prion protein reveals a mechanism for oligomerization. Nat Struct Biol 2001; 8:770-4.
40. Janowski R, Kozak M, Jankowska E et al. Human cystatin C, an amyloidogenic protein, dimerizes through three-dimensional domain swapping. Nat Struct Biol 2001; 8:316-320.
41. Nelson R, Eisenberg D. Recent atomic models of amyloid fibril structure. Curr Opin Struct Biol 2006; 16:260-5.
42. Nelson R, Sawaya MR, Balbirnie M et al. Structure of the cross-beta spine of amyloid-like fibrils. Nature 2005; 435:773-8.
43. Ivanova MI, Sawaya MR, Gingery M et al. An amyloid-forming segment of beta2-microglobulin suggests a molecular model for the fibril. Proc Natl Acad Sci USA 2004; 101:10584-9.
44. Sambashivan S, Liu Y, Sawaya MR et al. Amyloid-like fibrils of ribonuclease A with three-dimensional domain-swapped and native-like structure. Nature 2005; 437:266-9.
45. Staniforth RA, Giannini S, Higgins LD et al. Three-dimensional domain swapping in the folded and molten-globule states of cystatins, an amyloid-forming structural superfamily. EMBO Journal 2001; 20:4774-4781.
46. Eakin CM, Attenello FJ, Morgan CJ et al. Oligomeric assembly of native-like precursors precedes amyloid formation by beta-2 microglobulin. Biochemistry 2004; 43:7808-15.
47. Murray AJ, Head JG, Barker JJ et al. Engineering an intertwined form of CD2 for stability and assembly. Nature Structural Biology 1998; 5:778-782.
48. Spinelli S, Desmyter A, Frenken et al. Domain swapping of a llama VHH domain builds a crystal-wide beta-sheet structure. FEBS Lett 2004; 564:35-40.
49. Guo Z, Eisenberg D. Runaway domain swapping in amyloid-like fibrils of T7 endonuclease I. Proc Natl Acad Sci USA 2006; 103:8042-7.
50. Lawson CL, Benoff B, Berger T et al. coli trp repressor forms a domain-swapped array in aqueous alcohol. Structure 2004; 12:1099-108.

51. Zahn R, Liu A, Luhrs T et al. NMR solution structure of the human prion protein. Proc Natl Acad Sci USA 2000; 97:145-50.
52. Lee S, Eisenberg D. Seeded conversion of recombinant prion protein to a disulfide-bonded oligomer by a reduction-oxidation process. Nat Struct Biol 2003; 10:725-30.
53. Louis JM, Byeon IJ, Baxa U et al. The GB1 amyloid fibril: recruitment of the peripheral beta-strands of the domain swapped dimer into the polymeric interface. J Mol Biol 2005; 348:687-98.
54. Rocchi A, Pellegrini S, Siciliano G et al. Causative and susceptibility genes for Alzheimer's disease: a review. Brain Res Bull 2003; 61:1-24.
55. Choudhury D, Thompson A, Stojanoff V et al. X-ray structure of the FimC-FimH chaperone-adhesin complex from uropathogenic Escherichia coli. Science 1999; 285:1061-6.
56. Sauer FG, Futterer K, Pinkner JS et al. Structural basis of chaperone function and pilus biogenesis. Science 1999; 285:1058-61.
57. Remaut H, Waksman G. Structural biology of bacterial pathogenesis. Curr Opin Struct Biol 2004; 14:161-70.
58. Huntington JA, Read RJ, Carrell RW. Structure of a serpin-protease complex shows inhibition by deformation. Nature 2000; 407:923-6.
59. Lomas DA, Evans DL, Finch JT et al. The mechanism of Z alpha 1-antitrypsin accumulation in the liver. Nature 1992; 357:605-7.
60. Stein PE, Carrell RW. What do dysfunctional serpins tell us about molecular mobility and disease? Nat Struct Biol 1995; 2:96-113.
61. Sharp HL, Bridges RA, Krivit W et al. Cirrhosis associated with alpha-1-antitrypsin deficiency: a previously unrecognized inherited disorder. J Lab Clin Med 1969; 73:934-9.
62. Dafforn TR, Mahadeva R, Elliott PR et al. A kinetic mechanism for the polymerization of alpha1-antitrypsin. J Biol Chem 1999; 274:9548-55.
63. Davis RL, Holohan PD, Shrimpton AE et al. Familial encephalopathy with neuroserpin inclusion bodies. Am J Pathol 1999; 155:1901-13.
64. Davis RL, Shrimpton AE, Holohan PD et al. Familial dementia caused by polymerization of mutant neuroserpin. Nature 1999; 401:376-9.
65. Belorgey D, Crowther DC, Mahadeva R et al. Mutant Neuroserpin (S49P) that causes familial encephalopathy with neuroserpin inclusion bodies is a poor proteinase inhibitor and readily forms polymers in vitro. J Biol Chem 2002; 277:17367-73.
66. Alberts B. Molecular biology of the cell. 3rd ed. New York. Garland Pub 1994.
67. Schreuder HA, de Boer B, Dijkema R et al. The intact and cleaved human antithrombin III complex as a model for serpin-proteinase interactions. Nat Struct Biol 1994; 1:48-54.
68. Beauchamp NJ, Pike RN, Daly M et al. Antithrombins Wibble and Wobble (T85M/K): archetypal conformational diseases with in vivo latent-transition, thrombosis and heparin activation. Blood 1998; 92:2696-706.
69. Zhou A, Huntington JA, Carrell RW. Formation of the antithrombin heterodimer in vivo and the onset of thrombosis. Blood 1999; 94:3388-96.
70. Liu Y, Eisenberg D. 3D domain swapping: As domains continue to swap. Protein Sci 2002; 11:1285-99.
71. Sawaya MR, Guo S, Tabor S et al. Crystal structure of the helicase domain from the replicative helicase-primase of bacteriophage T7. Cell 1999; 99:167-77.
72. Gallagher T, Alexander P, Bryan P et al. Two crystal structures of the B1 immunoglobulin-binding domain of streptococcal protein G and comparison with NMR. Biochemistry 1994; 33:4721-9.
73. Shimba N, Kariya E, Tate S et al. Structural comparison between wild-type and P25S human cystatin A by NMR spectroscopy. Does this mutation affect the alpha-helix conformation? J Struct Funct Genomics 2000; 1:26-42.
74. Janowski R, Kozak M, Abrahamson M et al. 3D domain-swapped human cystatin C with amyloid-like intermolecular beta-sheets. Proteins 2005; 61:570-8.
75. Yamasaki M, Li W, Johnson DJ et al. Crystal structure of a stable dimer reveals the molecular basis of serpin polymerization. Nature 2008; 455(7217): 1255-8.
76. Kolodziejczyk R, Michalska K, Hernandez-Santoyo A et al. Crystal structure of human cystatin C stabilized against amyloid formation. FEBS J 2010; 277(7): 1726-37.
77. Yamasaki M, Sendall TJ, Pearce MC et al. Molecular basis of α1-antitrypsin deficiency revealed by the structure of a domain-swapped trimer. EMBO Rep 2011; 12(10): 1011-7.
78. Liu C, Sawaya MR, Eisenberg D. β_2-microglobulin forms three-dimensional domain-swapped amyloid fibrils with disulfide linkages. Nat Struct Mol Biol 2011; 18(1): 49-55.
79. Hafner-Bratkovic I, Bester R, Pristovsek P. Globular domain of the prion protein needs to be unlocked by domain swapping to support prion protein conversion. J Biol Chem 2011; 286(14): 12149-56.
80. Yang S, Cho SS, Levy Y et al. Domain swapping is a consequence of minimal frustration. Proc Natl Acad Sci U S A 2004; 101(38): 13786-91.

81. Levy Y, Cho SS, Onuchic JN, et al. A survey of flexible protein binding mechanisms and their transition states using native topology based energy landscapes. J Mol Biol 2005; 346(4): 1121-45.
82. Cho SS, Levy Y, Onuchic JN et al. Overcoming residual frustration in domain-swapping: the roles of disulfide bonds in dimerization and aggregation. Phys Biol 2005; 2(2): S44-55.
83. Yang S, Levine H, Onuchic JN et al. Structure of infectious prions: stabilization by domain swapping. FASEB J 2005; 19(13): 1778-82.
84. Yang S, Levine H, Onuchic JN. Protein oligomerization through domain swapping: role of inter-molecular interactions and protein concentration. J Mol Biol 2005; 352(1): 202-11.

CHAPTER 10

FROM ARTIFICIAL ANTIBODIES TO NANOSPRINGS

The Biophysical Properties of Repeat Proteins

Laura S. Itzhaki*,[1] and Alan R. Lowe[2]

[1]University of Cambridge, Department of Chemistry, Cambridge, UK; [2]Department of Physics, University of California-Berkeley, Berkeley, California, USA

**Corresponding Author: Laura S. Itzhaki—Email: lsi@hutchison-mrc.cam.ac.uk*

Abstract: In this chapter we review recent studies of repeat proteins, a class of proteins consisting of tandem arrays of small structural motifs that stack approximately linearly to produce elongated structures. We discuss the observation that, despite lacking the long-range tertiary interactions that are thought to be the hallmark of globular protein stability, repeat proteins can be as stable and as co-orperatively folded as their globular counterparts. The symmetry inherent in the structures of repeat arrays, however, means there can be many partly folded species (whether it be intermediates or transition states) that have similar stabilities. Consequently they do have distinct folding properties compared with globular proteins and these are manifest in their behaviour both at equilibrium and under kinetic conditions. Thus, when studying repeat proteins one appears to be probing a moving target: a relatively small perturbation, by mutation for example, can result in a shift to a different intermediate or transition state. The growing literature on these proteins illustrates how their modular architecture can be adapted to a remarkable array of biological and physical roles, both in vivo and in vitro. Further, their simple architecture makes them uniquely amenable to redesign—of their stability, folding and function—promising exciting possibilities for future research.

INTRODUCTION

Repeat proteins, such as ankyrin repeats, leucine rich repeats (LRR) and tetratricopeptide repeats (TPR), consist of tandem arrays of a small structural motif of ~20-40 amino acids that

Protein Dimerization and Oligomerization in Biology, edited by Jacqueline M. Matthews.

stack in a roughly linear fashion, creating elongated and superhelical architectures. They are mostly found in arrays of four, up to tens of, repeats and they can comprise an entire protein or a domain within a larger protein. Repeat protein structures have simple topologies composed entirely of short-range interactions between residues within a repeat or between adjacent repeats, in striking contrast to the more complex topologies of globular proteins that are stabilized by contacts between residues distant in sequence. They are ubiquitous and highly versatile, mediating molecular recognition in many different biological processes. A picture is emerging, from work carried out in the last five years, of the simplicity of this modular architecture which has enabled the following: (1) proteins containing consensus repeat motifs have been produced with enhanced thermodynamic stabilities compared with their natural counterparts; (2) designed consensus repeat proteins have been used successfully as scaffolds for engineering novel binding specificities; (3) repeat proteins have folding pathways that are amenable to design. None of these features have been found with any consistency for globular proteins. By far the most work has been carried out on ankyrin repeat proteins, followed by TPR proteins and these two will therefore be the focus of this chapter.

REPEAT PROTEIN STRUCTURES

As with globular proteins, the repeat protein family can be broken down further into sub-families based on their secondary structure content. A range of different repeat proteins are shown in Figure 1. The most common forms of α-helical repeat proteins are ankyrin repeats and TPR repeats. Ankyrin repeats (named after the protein Ankyrin in which these repeats were first identified[1]) have an L-shaped profile, with a long loop/β-hairpin sitting at right angles to a pair of antiparallel α-helices (Fig. 1A and B). The helices of one repeat pack against the helices of an adjacent repeat, with the interface between repeats consisting of mainly hydrophobic interactions stabilising the helices and hydrogen bonding stabilising the long loop region. The two helices of the protein are of different lengths causing curvature in the structure with each repeat rotated 2-3° counterclockwise relative to the previous one. The tetratricopeptide repeat (TPR) consists of a 34 residue motif[2,3] that adopts an antiparallel helix-turn-helix arrangement without any long connecting loops unlike ankyrin repeats. The packing of a pattern of small and large hydrophobic residues results in a rather splayed arrangement of the α-helices. The crystal structure of protein phosphatase 5 contains three TPR repeats and forms a right-handed superhelical structure with a continuous helical groove[4] (Fig. 1C). The groove is the proposed site of protein-protein interactions. The curvature that results from arrays of large numbers of repeats is particularly well-illustrated by the structure of the procine ribonuclease inhibitor[5] (Fig. 1G).

Interestingly, not all repeat proteins form linear elongated structures. The prokaryotic TPR protein N1p1 forms a globular structure in which repeats from the N-terminus form a central hydrophobic core and the C-terminal repeats form the outer solvent exposed surface, with the entire protein resembling a spiral.[6] Further, WD40 repeats form propeller-like structures with each repeat folding into a β-sheet resembling a blade of the propeller[7] (Fig. 1H).

REPEAT PROTEINS AS MEDIATORS IN MOLECULAR RECOGNITION

The large and regular surfaces of repeat proteins lend them ideally to mediating molecular recognition events. Repeat proteins provide a central scaffold with a conserved sequence motif and typically have highly variable sequences in loops presented on the

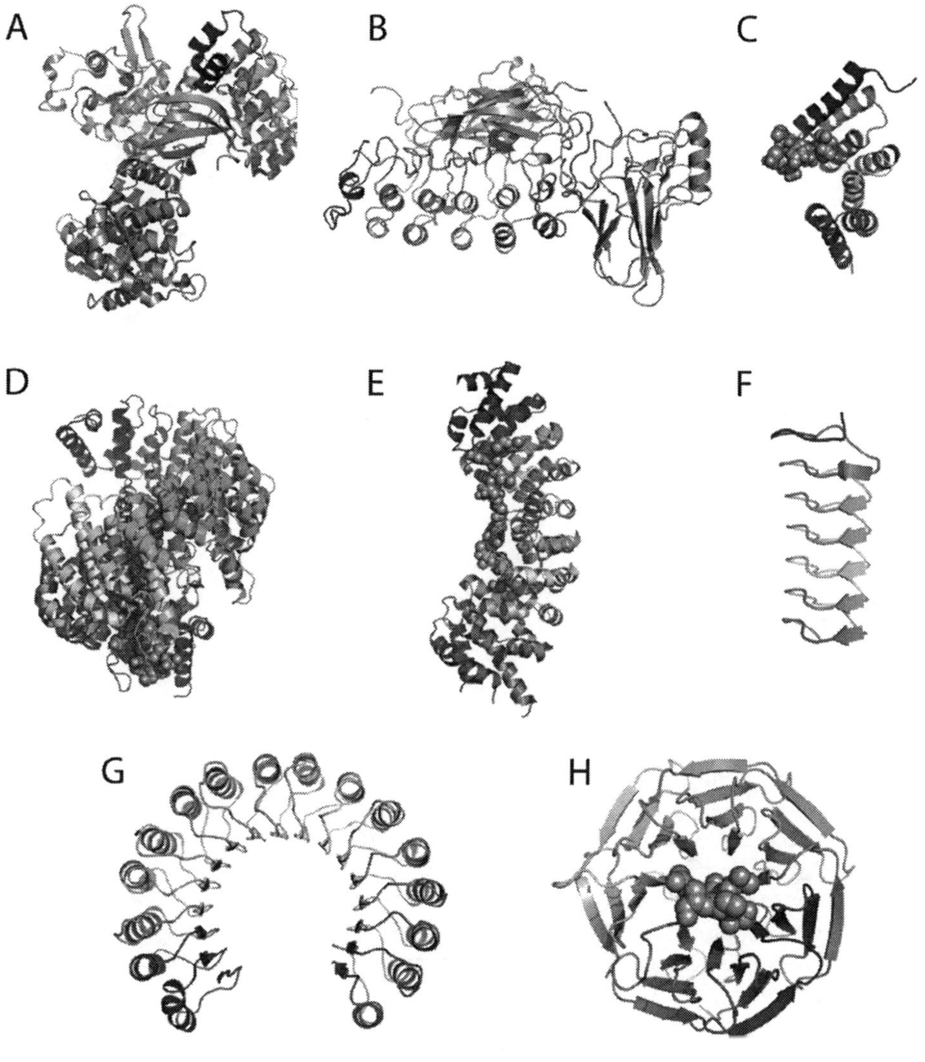

Figure 1. Repeat protein architectures. Interaction partners are colored in grey. A) ANK: the ternary complex between p18^{INK4c}, Cdk6 and K-cyclin (1G3N). B) ANK: The transcription factor complex between NFκB and IκBα (1IKN). C) TPR: The TPR domain from protein phosphatase 5 in complex with an Hsp90 derived peptide (2BUG). D) HEAT: Importin β interacting with the IBB peptide of Importin α (1QGK). E) ARM: Importin α in complex with the NLS peptide (1EE5). F) β-helix of an antifreeze protein from the *Tenebrio molitor* beetle (1EZG). G) LRR: The porcine ribonuclease inhibitor (2BNH). H) WD40: WDR5/Histone H3 complex (2H13). A color version of this figure is available online at www.landesbioscience.com/curie.

surface. In this way, the loops can provide varied interaction surfaces in terms of geometry as well as functional group content. Repeat proteins are frequently found in proteins containing other domains and they may serve to recruit substrates in these cases. An important class of such proteins are the F-box proteins.[8] SCF complexes are an important class of E3 ubiquitin ligases that target key cell cycle regulators and transcription factors to the proteasome for degradation. SCF complexes consist of two invariable, core subunits and a third, variable, subunit, the F-box protein that recruits substrates for ubiquitination by an associated E2 enzyme. F-box proteins have an F-box domain that binds to the core SCF subunits and a substrate recognition domain, in many cases composed of LRRs or WD40 repeats.

Ankyrin repeats have been extensively studied in terms of their protein-protein interactions. For example, the INK4 family of ankyrin repeat proteins specifically inhibit the cyclin-dependent kinases (CDK) of the cell cycle (Fig. 1A). There are four members of the INK4 family of CDK inhibitors: p15INK4B, p16INK4A, p18INK4C, p19INK4D. All four proteins are biochemically indistinguishable with respect to inhibition of CDK4/6. Understanding the mechanism of molecular recognition of these proteins is made more important by the fact that a common event in tumorigenesis is mutation of the regulatory proteins involved in the G1-S transition. The INK4 family of proteins specifically inhibit the cyclin-dependent kinases CDK4 and CDK6 responsible for initiating this transition. Insertions in the ankyrin repeat motif are common and usually found in the loop regions. In the case of the INK4 proteins, the first helix in the second repeat is shorter than the consensus one.

The ankyrin repeats of the protein IκBα are an interesting example of the phenomenon of protein folding induced by binding, as is seen in an increasing number of protein-protein interactions. IκBα is an inhibitor of the transcription factor NF-κB and the NF-κB-binding domain of IκBα has six ankyrin repeats (Fig. 1B). The structure of IκBα is only known in complex with NF-κB but extensive biophysical analysis revealed that although the protein is compactly folded and has all of its secondary structural elements it nevertheless has some highly dynamic, molten-globule-like regions in both free and complexed forms.[9] Subsequent analysis using hydrogen/deuterium exchange monitored by mass spectrometry showed that the β-hairpins of ankyrin repeats 5 and 6 only fold upon binding to NF-κB.[10,11] The authors propose that this feature of IκBα may be important for several aspects of its function that are critical for the tight regulation of NF-κB activity. In particular, the flexibility of the hairpins may help it to bind to different NF-κB dimers resulting in the transcription of different genes and also facilitate the dissociation of NF-κB from DNA.

Another example of the molecular recognition capabilities of repeat proteins is found in nuclear transport (reviewed in 12, 13). The import of cargos into the nucleus of cells is dependent on Armadillo repeat- and HEAT repeat-containing proteins known as the importins. The Armadillo repeat was first identified in the *Drosophila melanogaster* gene product Armadillo.[14] These proteins consist of tandem repeats of an approximately 40-residue motif which folds into a three-helix bundle. The approximately 40-residue HEAT motif consists of only two helices. The HEAT motif was initially found in a diverse range of proteins, including the four from which its name derives: huntingtin, elongation factor 3, the PR65/A subunit of protein phosphatase 2A and the lipid kinase Tor (target of rapamycin).[15] In both armadillo- and HEAT-repeat proteins, the motifs stack together to form an elongated super-helix. Cargoes destined for the nucleus and displaying the nuclear localisation signal (NLS),[16-19] a short predominantly positively charged peptide, bind importin-α, a protein containing 10 armadillo repeats in which the super-helix creates a narrow internal groove that envelops the NLS and displays its own positively charged peptide[20] (Fig. 1E). This importin-α peptide interacts with a 19-HEAT repeat protein, importin-β/β-karyopherin[21] (Fig. 1D). The resulting large complex enables the cargo to translocate through the nuclear pore. Not only are these interactions extremely specific for the NLS peptide, but the outer surface of these proteins also allow the complex to dissolve into the central channel of the pore itself in a highly selective fashion. A few cargo proteins bind directly to importin-β rather than via importin-α. Once in the nucleus, RanGTP binds to importin-β thereby dissociating the importin-αβ complex and leading to release of the cargo. Another 19-HEAT repeat protein, CAS, then exports importin-α out of the nucleus so that it can be used again. Crystal structures reveal that the HEAT repeats of both importin-β and CAS function as tightly wound springs with the curvature created by the packing of the individual α-helices and thereby the twist of the super-helix, changing dramatically depending on whether they are in the free or complexed form. It is thought that this flexibility enables them to bind to a range of different partners by wrapping around them with varying helicoidal pitch.

The regular presentation of surface residues provided by the repeat protein architecture lends itself to roles other than just mediating interactions between proteins. For example, the antifreeze proteins of some invertebrates form fascinating structures that are able to directly bind to ice crystals. These antifreeze proteins, found in invertebrates such as the *Tenebrio molitor* beetle, are composed of tandem 12-residue repeats (TCTxSxxCxxAx) which form linear beta-helical structures which are either square or triangular in cross section[22] (Fig. 1F). Remarkably, these beta-helical structures present a regular array of optimally orientated binding pockets which bind water molecules at the edges of ice crystals, thereby preventing further ice crystal growth and therefore damage to the organism.

DESIGNING REPEAT PROTEINS

A number of groups have taken advantage of the symmetry in linearly repeating structures to successfully design novel proteins that have been found to have increased stability relative to the natural counterparts. To date, they have all applied, at least in part, the concept of consensus design[23]—engineering a protein to have the most common residue as each position as determined by multiple sequence alignment. The rationale is that the residues that are most highly conserved among proteins of a particular fold will be those that are important for the stability of that fold, although there are a number of reasons why this rationale may be imperfect.[24] Two groups have designed consensus ankyrin repeat proteins, making use of the relatively high sequence homology within ankyrin repeats. Pluckthun's group used libraries assembled from N- and C-terminal capping repeats and an internal repeat consensus sequence with 27 fixed positions and the remaining 6 randomized. Six of the library members were chosen at random, containing between four and six repeats.[25-27] Crystallography of one of the proteins showed that it adopted the correct fold.[28,29] The proteins were more stable (as measured by equilibrium chemical denaturation experiments) than natural ankyrin repeat proteins of similar sizes and five of the six were monomeric. The most stable protein had six repeats and a free energy of unfolding that was greater than 20 kcal.mol^{-1}.[25] A comparison of the hydrophobic core packing of the designed ankyrin repeat structure with natural ankyrin repeats did not reveal any significant differences. The authors suggest that elimination of insertions, commonly found in the loop regions and used for molecular recognition, may contribute to the enhanced stability of these designed proteins relative to their natural counterparts; the loop regions form a greatly improved, regular network of hydrogen bonds. Increased stability is also likely to come from the conserved TPLH motif in the first helix of the ankyrin repeat which again results in a network of hydrogen bonds extending throughout the molecule, as well as from the tight packing of hydrophobic residues between the α-helices. A similar approach was used to design leucine-rich repeat protein libraries. As for the consensus ankyrin repeat proteins, the leucine-rich repeat proteins were found to be expressed at very high levels in *E.* coli, soluble, monomeric and stable.[30] Peng's group designed a consensus ankyrin repeat sequence that differed in four positions from Pluckthun's[31]. The other major difference was that they did not use capping repeats that were different from the internal repeats. They built proteins consisting of 1-4 ankyrin repeats and X-ray crystal structures of the two larger proteins showed that they adopted the correct fold.[31] The designed proteins were also found to have high thermal stability but they were only soluble in acidic conditions. The solubility at neutral pH was improved by making substitutions of surface leucine residues for arginines.

Pluckthun's group have since gone on to show that the idea of using consensus ankyrin repeats as scaffolds on which to graft novel binding specificities can have great success. The residues to be used for molecular recognition are located on the β-turn and first helix of each repeat, creating a large and modular surface. Noteworthy are designed 4-ankyrin repeat proteins (DARPins) selected against human epidermal growth factor receptor 2 with high specificity and low nanomolar affinities.[32] The interaction was subsequently improved to a 90 nM affinity using error-prone PCR and stringent off-rate selection.[33]

Regan's group have made novel TPR proteins by consensus design. They constructed proteins containing 1.5-3.5 TPR motifs[34] and additionally incorporated an N-capping, helix-stabilising GNS sequence at the N-terminus and an extra solvating helix at the C-terminus. The rationale for the latter addition was that similar helices were present in all of the structures of TPR-containing proteins and were found to increase solubility. The 2.5- and 3.5-TPR repeat proteins folded into the correct structures and had high thermal stabilities.[35] Whereas the enhanced stability of the consensus-designed ankyrin repeat proteins appears to come from the regularized hydrogen bonding networks, the stabilising interactions in the consensus TPR structures are predominantly the large hydrophobic residues that force the α-helices apart into the characteristic elongated architecture. This group then took the consensus TPR repeats and redesigned them to bind to the C-terminal peptide of the molecular chaperone Hsp90 (a natural TPR binding partner).[36] The designed protein had greater specificity than the natural TPR proteins.

Several studies have recently shown that the stabilities of naturally occurring ankyrin repeat proteins can be enhanced by using the consensus sequence.[37-39] For example, for the 7-ankyrin repeat domain of Notch, replacement of the two C-terminal repeats by consensus repeats increased the stability of the protein by almost 6 $kcal.mol^{-1}$.[38] Substitution by consensus residues of two residues in the N-terminal repeat of the 4-ankyrin repeat protein myotrophin increased the stability by over 2 $kcal.mol^{-1}$.[39] The 6-ankyrin repeat region of IkBa has only marginal stability; engineering of three residues, one in each of three of the repeats, to conform to the extended motif GXTPLHLA of the consensus (that creates a hydrogen bonding network as described above) resulted in an increase in stability of 4.5 kcal mol^{-1}.[37]

BIOPHYSICAL PROPERTIES OF REPEAT PROTEINS

The distinctive architecture of repeat proteins compared with globular proteins makes them a particular interesting target for biophysical analysis. The simple, modular nature of the structures and the lack of long-range interactions (which are thought to be contribute to the cooperative folding of globular proteins), pose a number of questions.

First, are repeat proteins more or less stable than globular proteins of similar size? Second, are repeat structures less cooperatively folded given the lack of long-range contacts? Third, are the differences in the structures of repeat proteins compared with globular proteins reflected in their folding mechanisms? In particular, it has been shown that there is a correlation between folding rate and proportion of short-range contacts in the native structure because there is a smaller entropic cost of closing a short loop to form an interaction between residues close in sequence than closing a long loop between sequence-distant residues;[40] so, do repeat proteins fold faster than globular ones because of the predominance of short-range contacts? And, are there multiple folding pathways accessible due to the linear symmetry of their structures?

Repeat Protein Stability

Perhaps surprisingly given the lack of long-range interactions, the natural ankyrin repeat proteins and designed TPR proteins studied to-date have stabilities within the range found for globular proteins. Several studies have shown that both the free energy of unfolding and the *m*-value (a measure of the size of the structural unit that is unfolding in a cooperative step) increase with increasing number of repeats, indicating that there is cooperativity of folding, although cooperativity does break down after a certain number of repeats is reached (see below).[25,35] Truncations of p16[41] and the Notch[42] ankyrin domain indicate that a minimum of two ankyrin repeats is required for a cooperative unit of structure, whereas Regan and colleagues were able to design a cooperatively folded polypeptide consisting of only one and a half TPR motifs. These findings confirm the observation from contact maps of ankyrin repeat and TPR proteins indicating that the TPR module is a more independently folded unit of structure that makes more contacts within the module than between modules, whereas the ankyrin module makes many contacts with adjacent modules.[43] In nature, ankyrin repeat and TPR-containing domains rarely contain fewer than three tandem motifs.

The biophysical properties of proteins consisting of 4-7 ankyrin repeats have been studied the most extensively. The data show that, despite their modular architecture, these ankyrin repeat proteins unfold in as cooperative a manner as do globular proteins. Studies of deletion variants of the Notch ankyrin domain suggest that cooperativity arises because the interaction between neighbouring repeats is highly stabilising whereas the interactions made by residues within a repeat are destabilising.[44] The observed cooperativity is particular impressive when one considers that the 7-ankyrin repeat proteins Notch and gankyrin each comprise over 200 residues and yet they appear to unfold in a two-state manner. From the globular proteins studied to-date, two-state unfolding appears to break down well before this size is reached. The less well-studied repeat proteins containing β-strand structure have also been found to unfold cooperativity,[45] although a series of designed LRR proteins of increasing length showed deviation from two-state behaviour[30] and unfolding of the very large, β-helical protein pertactin also occurred via a clearly observed intermediate.[46]

One feature of repeat proteins that does distinguish them from globular proteins and that arises from the modularity of their structures, is the capacity to make them bigger or smaller by simply inserting or deleting modules. Whereas the insertion of large structural elements into globular proteins would be expected to be high destabilising, this is not the case for repeat proteins. These types of changes have been made most comprehensively to the Notch ankyrin domain.[38,42,44,47] The variants in which an internal repeat was duplicated were somewhat more stable than the wild type, but a breakdown in cooperativity of unfolding was observed when more than one duplicated internal repeat was inserted. The inter-repeat interfaces were not optimized in these variants which could explain their reduced cooperativity. However, other results suggest an alternative explanation. First, the consensus ankyrin repeat proteins designed by Pluckthun and Peng did not explicitly include a consideration of inter-repeat packing, but these proteins did show two-state unfolding. It is likely, therefore, that the consensus sequence is optimized for inter-repeat stability, as is borne out by the crystal structures of the consensus repeat proteins which reveal extended H-bonding networks. Further, recent studies of Notch have shown that two-state behaviour is maintained when two of its repeats are replaced by consensus repeats, although when the two consensus repeats are added to the protein then cooperativity breaks down.[38]

These various studies indicate that cooperative unfolding of ankyrin repeat proteins (and probably other repeat proteins also) is governed predominantly by the number of repeats in the array. In addition to this primary factor, there will be more subtle, sequence-dependent effects. Thus, both the 7-ankyrin repeat proteins Notch[47,48] and gankyrin (R. Hutton and LSI, manuscript in preparation) show a break-down in cooperativity on mutation at certain sites but not at others. Both proteins unfold in a non-two-state manner when mutations are made in the two C-terminal repeats, but not in the five N-terminal repeats. The results indicate that stability is distributed unevenly between the repeats of a protein (as expected given the non-identity of repeat sequences in a natural protein) and suggest that an even energy distribution is required for cooperativity.[47]

Ising Models and Repeat Protein Folding

The inter-repeat coupling in the context of cooperativity has also been explored using one dimensional Ising models.[49] In these models, each repeat is represented as having one of two states or *spins* ($s_i = \pm 1$) which correspond to that repeat being either folded or unfolded. Further, each repeat has an intrinsic equilibrium constant for folding and there is an additional equilibrium constant which describes the interactions or coupling between adjacent repeats. Mello and Barrick applied this method to the equilibrium unfolding of Notch ankyrin domain, utilising values for the equilibrium constants derived from a series of variants having deletions of the internal repeats. Their analysis indicated that fully folded and fully unfolded conformations comprised the majority of all species across the entire denaturant range used in the experiment. Even at the midpoint of the unfolding transition, a maximum of only 2% of all conformations were partly folded molecules with less than 6 of the 7 repeats folded.[44] Again, these results provide compelling evidence for the high level of coupling between adjacent repeats in these natural structures.

In contrast, Kajander et al utilized the one dimensional Ising model to study the cooperativity of folding in designed TPR repeat proteins. Here, several repeat proteins with varying numbers of consensus repeat domains were synthesized. By simultaneously fitting the data for all of the proteins with varying numbers of TPR repeats, Kajander et al were able to extract the equilibrium constants for folding and also coupling. For TPR proteins containing multiple copies of the consensus sequence, it appears that the probability of partially unfolded conformations at the unfolding transition midpoint is high; up to 25% of the population was found to be in a partially unfolded state for a construct containing three repeats.[50] These results indicate that there may be some optimisation of the coupling for cooperative folding in naturally occurring proteins which is absent in proteins containing multiple repeats of an identical sequence.

Characterisation of Equilibrium (Un)Folding Intermediates

The studies of Notch, involving a series of variants having deletions of the internal repeats, indicate that the protein folds at equilibrium through an intermediate comprising folded central repeats and the terminal repeats unfolded.[44] Although comprising only five ankyrin repeats, the protein p19 unfolds at equilibrium via an intermediate which was recently characterized by NMR-monitored hydrogen/deuterium exchange.[51] This intermediate was found to have the three C-terminal repeats folded and the two N-terminal repeats unfolded.

Computer simulations of ankyrin repeats have made a number of predictions about their properties, including that the cooperativity breaks down when the protein contains more than six or seven repeats.[52] The sensitivity of the two-state unfolding of 7-ankyrin repeat proteins Notch and gankyrin to single amino acid substitutions bears out this prediction. However, recent studies of the much larger, 12-ankyrin repeat protein, D34, suggest that the potential degree of cooperativity in ankyrin repeat proteins could be much greater than was first envisaged. Wild-type D34 unfolds in two parts via an intermediate in which the N-terminal approximately six repeats are unfolded and the C-terminal six repeats are folded, a result that concurs with the idea that a cooperatively folded ankyrin 'domain' consists of six or seven repeats.[53] However, mutants can be made in which as many as eleven ankyrin repeats unfold in a single step. Mutations in the C-terminal repeats destabilize the wild-type intermediate, causing different intermediates to be populated. The closer to the C-terminus the mutation, the fewer repeats are structured in the intermediate; thus, structure in the intermediate frays from the site of the mutation, a domino-like effect.

Mechanisms of Repeat Protein Folding

As described above for the equilibrium behaviour, the basic kinetic folding characteristics of ankyrin repeat proteins appear also to mirror those of globular proteins. Thus, the refolding kinetics of myotrophin,[39,54] p16,[55,56] Notch ankyrin domain,[57]gankyrin (R. Hutton and LSI, manuscript in preparation) and p19,[51,58] comprising between 118 and 238 residues, are multi-phasic when monitored by CD or fluorescence and proceed via partly folded intermediates. Only for the designed 3-ankyrin repeat protein E1_5 were the refolding and unfolding kinetics monophasic.[59] The kinetic folding or unfolding mechanisms of five ankyrin repeat proteins have been studied in detail using a protein engineering approach. These are p16,[56] Notch ankyrin domain,[60] myotrophin,[39] D34 (N. Werbeck and LSI, submitted) and gankyrin (R. Hutton and LSI, manuscript in preparation). In the first of these studies, on p16, eighteen mutations spread throughout the four repeats showed a strikingly clear pattern of unfolding and refolding behaviour which indicated a transition state for unfolding with structure polarized in the C-terminal repeats.[56] Subsequent studies of myotrophin and gankyrin also indicated a structurally polarized folding mechanism.[39] However, a different picture emerged for the Notch ankyrin domain. Analysis of seven mutants, one in each repeat, showed that folding is initiated at three internal repeats.[60] On the one hand, the terminal repeats might be expected to fold first since the entropic cost of fixing the end of the polypeptide chain would be less than for fixing the internal repeats. On the other hand the internal repeats might be expected to fold first since these have the potential to make stabilizing interactions with more neighbouring repeats than the terminal repeats. However, it is likely that the fine details of each protein sequence also play a critical role. For globular proteins, there is a balance between the entropic cost of closing a loop in order to bring distant residues together and the enthalpic gain of the resulting contacts that are made. The region of a globular protein that folds first is therefore the one that uses the best contacts for which the entropy penalty is minimal. The linear symmetry of repeat proteins, however, means that equivalent positions in each repeat of a protein pay the same entropic cost of bringing other residues into contact with it. It is only the enthalpic gain that varies between repeats, being determined by the individual sequences of the different repeats. Therefore, the repeat(s) that folds first will be the ones that are the most stable, assuming that the lowest energy folding route (i.e., transition state) corresponds to the lowest energy folded repeat(s). This simple picture of folding the most stable repeats first is borne out by experiments, described below, that show the folding pathway can be changed in a predictable way simply by manipulating the stabilities of individual repeats. Finally, for each of p16, myotrophin, Notch and gankyrin, the structure of the rate-determining transition states comprises more than a single ankyrin motif. This result is as expected, since the ankyrin motif in isolation has been shown to be intrinsically unstable whereas the interaction between motifs is highly favourable.

Multiple Folding Pathways of Ankyrin Repeat Proteins

The most striking feature of both myotrophin and gankyrin (and also D34, see below) is that there are alternative pathways accessible to the native state in which folding is initiated at one or other end of the structure.[39] The authors were able to redesign the folding pathway of myotrophin very simply. By taking advantage of the modular structure and manipulating the stabilities of the individual repeats, they could switch the folding initiation site from one end of the structure to the other.[39] Molecular dynamics simulations carried out on a number of ankyrin repeat proteins, before the experimental data were available, also predicted pathway heterogeneity.[61] The concept of energy landscapes suggests that a protein can follow multiple folding pathways. However, there has been very little experimental evidence to support this view for globular proteins. Experimental studies indicate that folding transition states of small globular proteins are represented by relatively homogenous ensembles of structures (excluding parallel folding reactions that result from heterogeneity in the denatured state, due to proline isomerisation for example). Moreover, the folding mechanisms of globular proteins are generally robust and therefore only a drastic change in the energetic balance, by circular permutation for example, can in some cases shift the folding nucleus from one part of the structure to another. By contrast, the potential to initiate folding at more than one site may be a general feature of repeat proteins that arises from the symmetry inherent in their structures.

Alternative pathways were, however, not observed for p16 or Notch (although a Notch variant containing two consensus-stabilized repeats showed evidence of a shift to an alternative folding pathway[38]). Presumably the different behaviour reflects the distribution of stability across the repeats. Thus, when the repeats within a protein have similar stabilities then multiple pathways may be accessible if the energy barriers to their folding are also of similar energy; by contrast, when some repeats have significantly greater stabilities than others then there is a unique folding pathway. Interestingly, preliminary studies on the 12-ankyrin repeat D34 indicate that the two halves of the protein each display unfolding mechanisms resembling one or other of these two scenarios (N. Werbeck and LSI, manuscript submitted). The C-terminal half, which folds and unfolds in the absence of a folded N-terminal half, can follow alternative pathways; in contrast, the N-terminal half folds and unfolds in the presence of a folded C-terminal half which therefore acts as a "seed" and consequently directs folding along a unique pathway.

Mechanical Properties of Ankyrin Repeats

In addition to the solution folding properties of repeat proteins, there has been recent interest in their folding and unfolding under mechanical perturbation. Whereas the curvature created in smaller repeat proteins provides a concave face that is ideal as an interaction surface, structures composed of a greater number of repeats resemble solenoids, forming a continuous spiral or spring-like topology. Large ankyrin repeat proteins consisting of many tens of repeats have been postulated to have mechano-signal transduction roles in processes such as hearing. Stacks of ankyrin repeats found in the cytoplasmic domains of the transient receptor potential (TRP) ion channels have been proposed as springs which provide a resting tension.[62-64] These ion channels are mechanically gated by deflections of the stereocillia in response to sound. Recent work using single molecule atomic force microscopy (AFM) has demonstrated that ankyrin repeat proteins work as "nanosprings" within this role; their modular nature enables them to behave as a reversible spring that is capable of generating force upon refolding.[65] Further, the repeats unfold sequentially under mechanical perturbation.[66] These observations are consistent with those from the solution studies of folding and demonstrate the remarkable range of cellular roles of repeat proteins.

Self-Assembly and Higher-Order Structures in Repeat Proteins

One of the most interesting new classes of repeat proteins to emerge in recent years is the reflectin family of proteins, isolated from the Hawaiian bobtail squid. Although no structural information is currently available, the proteins are known to contain five repeats of a 18-20-residue sequence motif [M/FD(X)5MD(X)5MD(X)3/4] which is unusually rich in methionine.[67] These proteins have the remarkable capability of self-assembling into regular higher-order structures that act as biological diffraction gratings, reflecting light in the visible wavelengths.[68] Large platelets of these proteins form photonic structures which can be used to modulate macroscale changes such as overall body colouration in the squid. These proteins are providing a unique challenge to biophysicists to understand their folding and assembly as well as their physical properties.

CONCLUSION

Repeat proteins have turned out to be a fascinating class of structures, with particular appeal to those interested in protein folding, engineering and design. They are modular and therefore much more amenable to redesign than are globular proteins and yet this feature does not appear to compromize cooperativity, a hallmark of globular proteins that is thought to be important in making them sufficiently robust to maintain them in their native structures over their functional lifetimes. The recent studies reviewed here reveal 'design ability' in every aspect of repeat proteins, including their stability, folding and function. In the last few years, repeat proteins of ever increasing size and complexity have been tackled, and these are providing a new awareness of how folding regulates function in repeat proteins.[69-80] In parallel, computational analysis of repeat proteins has provided thought-provoking insights into the distinct properties of these one-dimensional structures when compared with globular proteins.[81-84] With this body of experimental and theoretical work allowing us to understand how repeat proteins behave and how to build them from their constituent parts, we are now well-placed to design artificial repeat proteins and thereby to fully exploit their potential in biotechnology—for therapeutic and diagnostic applications, as biosensors and as building blocks for new biomaterials.[85-88]

REFERENCES

1. Bennett V, Stenbuck PJ. Identification and partial purification of ankyrin, the high affinity membrane attachment site for human erythrocyte spectrin. J Biol Chem 1979; 254(7):2533-41.
2. Hirano T, Kinoshita N, Morikawa K et al. Snap helix with knob and hole: essential repeats in S. pombe nuclear protein nuc2+. Cell 1990; 60(2):319-28.
3. Sikorski RS, Boguski MS, Goebl M et al. A repeating amino acid motif in CDC23 defines a family of proteins and a new relationship among genes required for mitosis and RNA synthesis. Cell 1990; 60(2):307-17.
4. Das AK, Cohen PW, Barford D. The structure of the tetratricopeptide repeats of protein phosphatase 5: implications for TPR-mediated protein-protein interactions. EMBO J 1998; 17(5):1192-9.
5. Kobe B, Deisenhofer J. Crystal structure of porcine ribonuclease inhibitor, a protein with leucine-rich repeats. Nature 1993; 366(6457):751-6.
6. Wilson CG, Kajander T, Regan L. The crystal structure of NlpI. A prokaryotic tetratricopeptide repeat protein with a globular fold. Febs J 2005; 272(1):166-79.
7. Jawad Z, Paoli M. Novel sequences propel familiar folds. Structure 2002; 10(4): 447-454.
8. Willems AR, Goh T, Taylor L et al. SCF ubiquitin protein ligases and phosphorylation-dependent proteolysis. Philos Trans R Soc Lond B Biol Sci 1999; 354(1389):1533-50.
9. Croy CH, Bergqvist S, Huxford T et al. Biophysical characterization of the free IkappaBalpha ankyrin repeat domain in solution. Protein Sci 2004; 13(7):1767-77.
10. Bergqvist S, Croy CH, Kjaergaard M et al. Thermodynamics reveal that helix four in the NLS of NF-kappaB p65 anchors IkappaBalpha, forming a very stable complex. J Mol Biol 2006; 360(2):421-34.
11. Truhlar SM, Torpey JW, Komives EA. Regions of IkappaBalpha that are critical for its inhibition of NF-kappaB. DNA interaction fold upon binding to NF-kappaB. Proc Natl Acad Sci USA 2006; 103(50):18951-6.
12. Conti E, Muller CW, Stewart M. Karyopherin flexibility in nucleocytoplasmic transport. Curr Opin Struct Biol 2006; 16(2):237-44.
13. Stewart M, Molecular mechanism of the nuclear protein import cycle. Nat Rev Mol Cell Biol 2007; 8(3):195-208.
14. Riggleman B, Wieschaus E, Schedl P. Molecular analysis of the armadillo locus: uniformly distributed transcripts and a protein with novel internal repeats are associated with a Drosophila segment polarity gene. Genes Dev 1989; 3(1):96-113.
15. Andrade MA, Bork P. HEAT repeats in the Huntington's disease protein. Nat Genet 1995; 11(2):115-6.
16. Chi NC, Adam EJH, Adam SA. Sequence and characterization of cytoplasmic nuclear-protein import factor P97. J Cell Biol 1995; 130(2):265-274.
17. Gorlich D, Kostka S, Kraft R. et al. 2 Different subunits of importin cooperate to recognize nuclear-localization signals and bind them to the nuclear-envelope. Curr Biol 1995; 5(4):383-392.
18. Imamoto N, Shimamoto T, Kose S et al. The nuclear pore-targeting complex binds to nuclear-pores after association with a karyophile. FEBS Lett 1995; 368(3):415-419.
19. Radu A, Blobel G, Moore MS. Identification of a protein complex that is required for nuclear-protein import and mediates docking of import substrate to distinct nucleoporins. Proc Natl Acad Sci U S A 1995; 92(5):1769-1773.
20. Andrade MA, Petosa C, O'Donoghue SI et al. Comparison of ARM and HEAT protein repeats. J Mol Biol 2001; 309(1):1-18.
21. Cingolani G, Petosa C, Weis K et al. Structure of importin-beta bound to tbe IBB domain of importin-alpha. Nature 1999; 399(6733):221-229.
22. Liou YC, Tocilj A, Davies PL et al. Mimicry of ice structure by surface hydroxyls and water of a beta-helix antifreeze protein. Nature 2000; 406(6793):322-4.
23. Desjarlais JR, Berg JM. Use of a zinc-finger consensus sequence framework and specificity rules to design specific DNA binding proteins. Proc Natl Acad Sci USA 1993; 90(6):2256-60.
24. Magliery TJ, Regan L. Beyond consensus: statistical free energies reveal hidden interactions in the design of a TPR motif. J Mol Biol 2004; 343(3):731-45.
25. Binz HK, Stumpp MT, Forrer P et al. Designing repeat proteins: well-expressed, soluble and stable proteins from combinatorial libraries of consensus ankyrin repeat proteins. J Mol Biol 2003; 332(2):489-503.
26. Forrer P, Stumpp MT, Binz HK et al. A novel strategy to design binding molecules harnessing the modular nature of repeat proteins. FEBS Lett 2003; 539(1-3):2-6.
27. Kohl A, Binz HK, Forrer P et al. Designed to be stable: crystal structure of a consensus ankyrin repeat protein. Proc Natl Acad Sci USA 2003; 100(4):1700-5.

28. Binz HK, Amstutz P, Kohl A et al. High-affinity binders selected from designed ankyrin repeat protein libraries. Nat Biotechnol 2004; 22(5):575-82.
29. Binz HK, Kohl A, Pluckthun A et al. Crystal structure of a consensus-designed ankyrin repeat protein: implications for stability. Proteins 2006; 65(2):280-4.
30. Stumpp MT, Forrer P, Binz HK et al. Designing repeat proteins: modular leucine-rich repeat protein libraries based on the mammalian ribonuclease inhibitor family. J Mol Biol 2003; 332(2):471-87.
31. Mosavi LK, Minor DL Jr, Peng ZY. Consensus-derived structural determinants of the ankyrin repeat motif. Proc Natl Acad Sci USA 2002; 99(25):16029-34.
32. Zahnd C, Pecorari F, Straumann N et al. Selection and characterization of Her2 binding-designed ankyrin repeat proteins. J Biol Chem 2006; 281(46):35167-75.
33. Zahnd C, Wyler E, Schwenk JM et al. A designed ankyrin repeat protein evolved to picomolar affinity to Her2. J Mol Biol 2007; 369(4):1015-28.
34. Main ER, Xiong Y, Cocco MJ et al. Design of stable alpha-helical arrays from an idealized TPR motif. Structure 2003; 11(5):497-508.
35. Main ER, Stott K, Jackson SE et al. Local and long-range stability in tandemly arrayed tetratricopeptide repeats. Proc Natl Acad Sci USA 2005; 102(16):5721-6.
36. Cortajarena AL, Kajander T, Pan W et al. Protein design to understand peptide ligand recognition by tetratricopeptide repeat proteins. Protein Eng Des Sel 2004; 17(4):399-409.
37. Ferreiro DU, Cervantes CF, Truhlar SM et al. Stabilizing IkappaBalpha by "consensus" design. J Mol Biol 2007; 365(4):1201-16.
38. Tripp KW, Barrick D. Enhancing the stability and folding rate of a repeat protein through the addition of consensus repeats. J Mol Biol 2007; 365(4):1187-200.
39. Lowe AR, Itzhaki LS. Rational redesign of the folding pathway of a modular protein. Proc Natl Acad Sci USA 2007; 104(8):2679-84.
40. Plaxco KW, Simons KT, Baker D. Contact order, transition state placement and the refolding rates of single domain proteins. J Mol Biol 1998; 277(4):985-94.
41. Zhang B, Peng Z. A minimum folding unit in the ankyrin repeat protein p16(INK4). J Mol Biol 2000; 299(4):1121-32.
42. Tripp KW, Barrick D. The tolerance of a modular protein to duplication and deletion of internal repeats. J Mol Biol 2004; 344(1):169-78.
43. Main ER, Jackson SE, Regan L. The folding and design of repeat proteins: reaching a consensus. Curr Opin Struct Biol 2003; 13(4):482-9.
44. Mello CC, Barrick D. An experimentally determined protein folding energy landscape. Proc Natl Acad Sci USA 2004; 101(39):14102-7.
45. Freiberg A, Machner MP, Pfeil W et al. Folding and stability of the leucine-rich repeat domain of internalin B from Listeri monocytogenes. J Mol Biol 2004; 337(2):453-61.
46. Junker M, Schuster CC, McDonnell AV et al. Pertactin beta-helix folding mechanism suggests common themes for the secretion and folding of autotransporter proteins. Proc Natl Acad Sci USA 2006; 103(13):4918-23.
47. Street TO, Bradley CM, Barrick D. Predicting coupling limits from an experimentally determined energy landscape. Proc Natl Acad Sci USA 2007; 104(12):4907-12.
48. Bradley CM, Barrick D. Limits of cooperativity in a structurally modular protein: response of the Notch ankyrin domain to analogous alanine substitutions in each repeat. J Mol Biol 2002; 324(2):373-86.
49. Zimm BH, Bragg JK. Theory of the phase transition between helix and random coil in polypeptide chains. J Chem Phys 1959; 31(2):526-535.
50. Kajander T, Cortajarena AL, Main ERG et al. A new folding paradigm for repeat proteins. J Am Chem Soc 2005; 127(29):10188-10190.
51. Low C, Weininger U, Zeeb M et al. Folding mechanism of an ankyrin repeat protein: scaffold and active site formation of human CDK inhibitor p19(INK4d). J Mol Biol 2007; 373(1):219-31.
52. Ferreiro DU, Cho SS, Komives EA et al. The energy landscape of modular repeat proteins: Topology determines folding mechanism in the ankyrin family. J Mol Biol 2005; 354(3):679-692.
53. Werbeck ND, Itzhaki LS. Probing a moving target with a plastic unfolding intermediate of an ankyrin-repeat protein. Proc Natl Acad Sci USA 2007; 104(19):7863-8.
54. Lowe AR, Itzhaki LS. Biophysical characterisation of the small ankyrin repeat protein myotrophin. J Mol Biol 2007; 365(4):1245-55.
55. Tang KS, Guralnick BJ, Wang WK et al. Stability and folding of the tumour suppressor protein p16. J Mol Biol 1999; 285(4):1869-1886.
56. Tang KS, Fersht AR, Itzhaki LS. Sequential unfolding of ankyrin repeats in tumor suppressor p16. Structure 2003; 11(1):67-73.
57. Mello CC, Bradley CM, Tripp KW et al. Experimental characterization of the folding kinetics of the notch ankyrin domain. J Mol Biol 2005; 352(2):266-81.
58. Zeeb M, Rosner H, Zeslawski W et al. Protein folding and stability of human CDK inhibitor p19(INK4d). J Mol Biol 2002; 315(3):447-457.

59. Devi VS, Binz HK, Stumpp MT et al. Folding of a designed simple ankyrin repeat protein. Protein Sci 2004; 13(11):2864-70.
60. Bradley CM, Barrick D. The notch ankyrin domain folds via a discrete, centralized pathway. Structure 2006; 14(8):1303-12.
61. Ferreiro DU, Cho SS, Komives EA et al. P versus Q: The energy landscape of modular repeat proteins: topology determines folding mechanism in the ankyrin family. J Mol Biol 2005; 354: 679-92.
62. Howard J, Bechstedt S. Hypothesis: A helix of ankyrin repeats of the NOMPIC-TRP ion channel is the gating spring of mechanoreceptors. Curr Biol 2004; 14(6):R224-R226.
63. Gillespie PG, Dumont RA, Kachar B. Have we found the tip link, transduction channel and gating spring of the hair cell? Curr Opin Neurobiol, 2005; 15(4):389-396.
64. Sotomayor M, Corey DP, Schulten K. In search of the hair-cell gating spring: Elastic properties of ankyrin and cadherin repeats. Structure 2005; 13(4):669-682.
65. Lee G, Abdi K, Jiang Y et al. Nanospring behaviour of ankyrin repeats. Nature 2006; 440(7081):246-249.
66. Li LW, Wetzel S, Pluckthun A et al. Stepwise unfolding of ankyrin repeats in a single protein revealed by atomic force microscopy. Biophys J 2006; 90(4):L30-L32.
67. Crookes WJ, Ding LL, Huang QL et al. Reflectins: The unusual proteins of squid reflective tissues. Science 2004; 303(5655):235-238.
68. Kramer RM, Crookes-Goodson WJ, Naik RR. The self-organizing properties of squid reflectin protein. Nat Mater 2007; 6(7):533-8.
69. Coleman ML, McDonough MA, Hewitson KS et al. Asparaginyl hydroxylation of the Notch ankyrin repeat domain by factor inhibiting hypoxia-inducible factor. J Biol Chem 2007; 282(33): 24027-38.
70. Wetzel SK, Settanni G, Kenig M, et al. Folding and unfolding mechanism of highly stable full-consensus ankyrin repeat proteins. J Mol Biol 2008; 376: 241-57.
71. Werbeck ND, Rowling PJ, Chellamuthu VR et al. Shifting transition states in the unfolding of a large ankyrin repeat protein. Proc Natl Acad Sci USA 2008; 105: 9982-7.
72. Evans MS, Sander IM, Clark PL. Cotranslational folding promotes beta-helix formation and avoids aggregation in vivo. J Mol Biol 2008; 383(3): 683-92.
73. Truhlar SM, Mathes E, Cervantes CF, et al. Pre-folding IkappaBalpha alters control of NF-kappaB signaling. J Mol Biol 2008; 380: 67-82.
74. Javadi Y, Main ER. Exploring the folding energy landscape of a series of designed consensus tetratricopeptide repeat proteins. Proc Natl Acad Sci U S A 2009; 106: 17383-8.
75. Kloss E, Barrick D. C-terminal deletion of leucine-rich repeats from YopM reveals a heterogeneous distribution of stability in a cooperatively folded protein. Protein Sci 2009; 18: 1948-60.
76. Forwood JK, Lange A, Zachariae U et al Quantitative structural analysis of importin-beta flexibility: Paradigm for solenoid protein structures. Structure 2010; 18: 1171-1183.
77. Low C, Homeyer N, Weininger U et al. Conformational switch upon phosphorylation: human CDK inhibitor p19INK4d between the native and partially folded state. ACS Chem Biol 2009; 4: 53-63.
78. Wetzel SK, Ewald C, Settani G et al. Residue-resolved stability of full-consensus ankyrin repeat proteins probed by NMR. J Mol Biol 2010; 402: 241-258.
79. Kim M, Abdi K, Lee G et al. Fast and forceful refolding of stretched alpha-helical solenoid proteins. Biophys J 2010; 98(12): 3086-92.
80. Serquera D, Lee W, Settanni G et al. Mechanical unfolding of an ankyrin repeat protein. Biophys J 2010; 98(7): 1294-301.
81. Ferreiro DU, Wolynes PG. The capillarity picture and the kinetics of one-dimensional protein folding. Proc Natl Acad Sci U S A 2008; 105: 9853-4.
82. Ferreiro DU, Walczak AM, Komives EA et al. The energy landscapes of repeat-containing proteins: topology, cooperativity, and the folding funnels of one-dimensional architectures. PLoS Comput Biol 2008; 4, e1000070.
83. Hagai T, Levy Y. Folding of elongated proteins: conventional or anomalous? J Am Chem Soc 2008; 130: 14253-62.
84. Grinthal A, Adamovic I, Weiner B et al. PR65, the HEAT-repeat scaffold of phosphatase PP2A, is an elastic connector that links force and catalysis. Proc Natl Acad Sci U S A 2010; 107: 2467-72.
85. Zahnd C, Kawe M, Stumpp MT et al. Efficient targeting with high affinity designed ankyrin repeat proteins: effects of affinity and molecular size. Cancer Res 2010; 70: 1595-1605.
86. Patricia MK, Stefan N, Rothschild S et al. A novel fusion toxin derived from an EpCAM-specific designed ankyrin repeat protien has potent antitumor activity. Clin Cancer Res 2011; 17: 100-110.
87. Cortajarena AL, Liu TY, Hochstrasser M et al. Designed proteins to modulate cellular networks. ACS Chem Biol 2010; 5: 545-552.
88. Grove TZ, Osuji CO, Forster JD, et al. Stimuli-responsive smart gels realized via modular protein design. J Am Chem Soc 2010; 132(40): 14024-6.

INDEX

L

M

N

O

P

Q

R

S

T

V

W

Y

Z